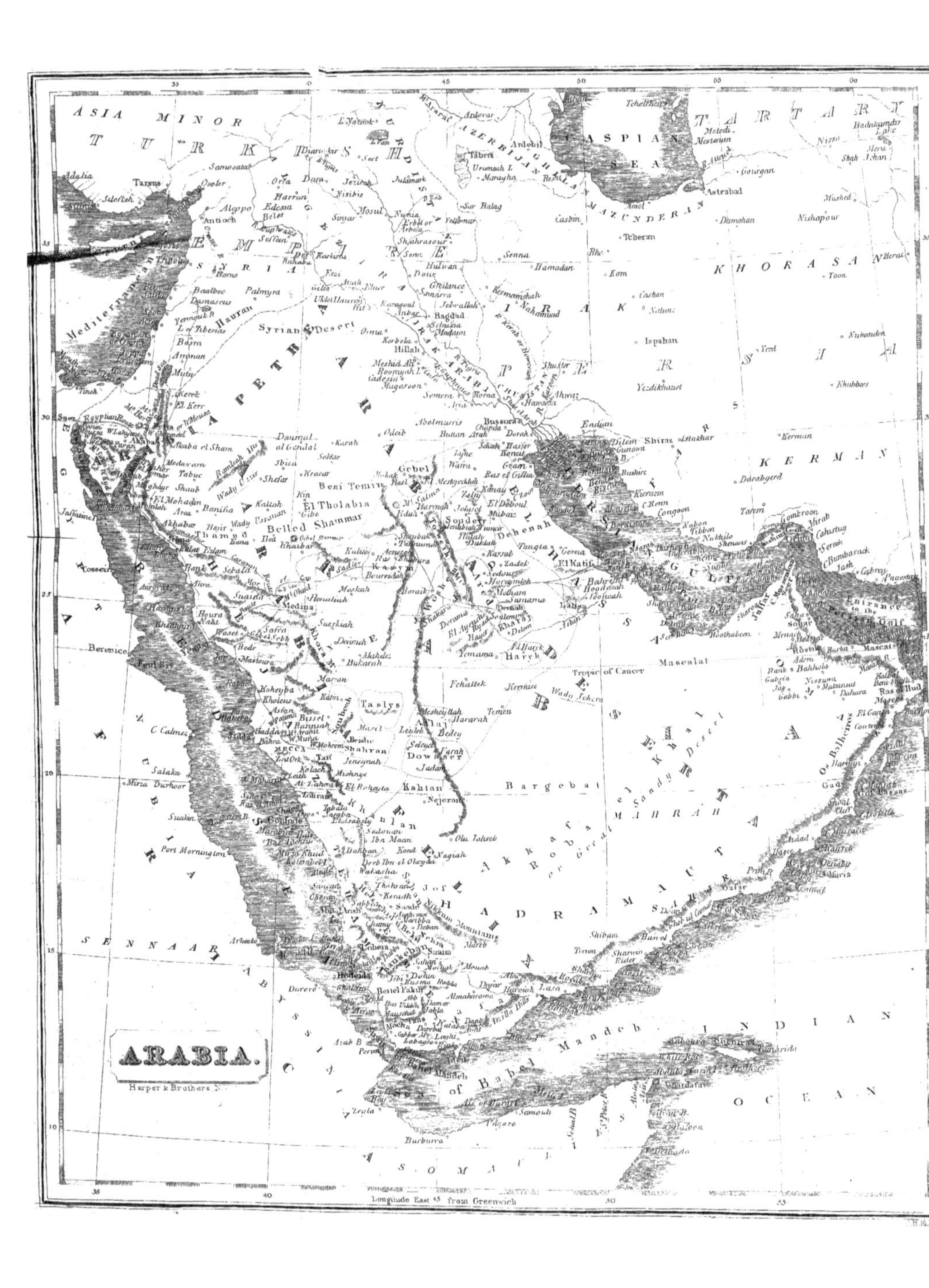
ARABIA.
Harper & Brothers
Longitude East 45 from Greenwich
ASIA MINOR
KHORASAN
KERMAN
CASPIAN SEA
SYRIA
Syrian Desert
Mediterranean
Belled Shammar
El Tholabia
Beni Temin
Medina
Mecca
Bargebat
Tropic of Cancer
INDIAN OCEAN
HADRAMAUT
Babel Mandeb
Bussora
Bagdad
Ispahan
Teheran
Aleppo
Antioch
Damascus
Palmyra
Hodeida
Port Mornington
Berenice
Masenlat

ARABIA.

VOL. I.

MOHAMMED.

HARPER & BROTHERS, NEW YORK.

THE

HISTORY OF ARABIA,

ANCIENT AND MODERN.

CONTAINING

A DESCRIPTION OF THE COUNTRY—AN ACCOUNT OF ITS INHABITANTS, ANTIQUITIES, POLITICAL CONDITION, AND EARLY COMMERCE—THE LIFE AND RELIGION OF MOHAMMED—THE CONQUESTS, ARTS, AND LITERATURE OF THE SARACENS—THE CALIPHS OF DAMASCUS, BAGDAD, AFRICA, AND SPAIN—THE CIVIL GOVERNMENT AND RELIGIOUS CEREMONIES OF THE MODERN ARABS—ORIGIN AND SUPPRESSION OF THE WAHABEES—THE INSTITUTIONS, CHARACTER, MANNERS, AND CUSTOMS OF THE BEDOUINS—AND A COMPREHENSIVE VIEW OF ITS NATURAL HISTORY.

BY ANDREW CRICHTON.

With a Map and Engravings.

IN TWO VOLUMES.

VOL. I

NEW YORK:
HARPER & BROTHERS, PUBLISHERS,
329 & 3 PEARL STREET,
FRANKLIN SQUARE
1868.

PREFACE.

It has been frequently remarked with surprise and regret, that while the annals of almost every nation of any political importance have been illustrated by British talent, no writer has hitherto favoured the world with a regular and continuous history of the Arabs. This neglect seems the more extraordinary in an age so distinguished as the present for literary enterprise, and when so many valuable accessions have been recently made to our scanty knowledge of the Arabian peninsula in the journals of intelligent travellers and scientific expeditions. Considering the many great and diversified events which the subject embraces, and the feelings of romantic interest that still attach to the celebrated regions of gold and frankincense, there appears some reason for the complaint that so little has been done to elucidate the character and actual condition of this ancient and renowned people, whose exploits once filled all Europe with astonishment; and that so much yet remains unknown of the sandy deserts they inhabit, and the singular institutions by which they are governed.

An attempt to supply this omission by connecting the records of the past with the illustrations of modern discovery, so as to exhibit the whole at one view and within a moderate compass, is the object of the following volumes. In entering upon a field so ample—unfolding in rapid succession a series of wars, revolutions, and vicissitudes of human fortune without parallel in any age or country—the author was not insensible of the numerous difficulties to be encountered. With what degree of success his labours have been attended remains for others to determine. At the same time, it is gratifying to reflect that at no former period could the task have been undertaken with so many facilities and advantages as at the present moment. The barriers of religious prejudice, which so long kept asunder the Christian and Mohammedan nations, are in a great measure broken down; the shades of ignorance and romance which in the infancy of navigation brooded over the people and the productions of Arabia have been dispelled; the character of the wandering Bedouin has been studied in his own deserts; even the Holy Land of Islam has been trodden by the feet of unbelievers, and the uncircumcised stranger has mingled in the sacred ceremonies of the Kaaba. These circumstances, by bringing to light many new and important facts, have furnished the historian with a rich stock of materials which a few years ago no European writer possessed. Of these sources of information the author has not neglected to avail himself; and, while acknowledging his obligations to the distinguished travellers, Niebuhr and Burckhardt, he ought also to state that he has not omitted to consult the more recent surveys of Chesney, Head, and Owen.

In the earlier chapters of the work, which refer to the dark and legendary times prior to the Mohammedan era, the author has endeavoured to give as clear and succinct an account of the primitive inhabitants, government, customs, and ancient commerce of the country, as the peculiar nature of his materials would admit. All historians and chronologists who have studied this obscure era have found themselves so bewildered with fable and tradition, or involved in such inextricable confusion from the want of authentic records, that they have been compelled either to rest satisfied with probable conjecture, or to abandon the subject in despair. That the author has succeeded in verifying doubts or reconciling anachronisms which perplexed the ablest Arabian antiquaries—Pococke, Reiske, and De Sacy—it would be presumption in him to assert. He has employed every means in his power, however, to discover the truth. For this purpose the oriental writers, Abulfeda, Tabiri, Masoudi, Hamza, Nuvairi, Abulfarage, and others who record the transactions of these remote ages, have been carefully perused; nor have those incidental notices and allusions been overlooked which occur in the pages of the Greek and Roman classics.

The life and religion of Mohammed form a curious and important episode in Arabian history; as giving rise to one of the most wonderful revolutions that the world has ever beheld. In treating of these, it has been the object of the present writer to give a fair representation of both without being swayed by any of those prejudices and apprehensions which have led some authors to speak of the character of that remarkable personage, and the institutions of which he was the founder, in a tone of such uncharitable rancour as to bring into suspicion the veracity of their statements. While shunning the bitter invectives of one class of biographers, he has avoided the panegyrical strain of others, who have endowed the apostle of the Koran with all the miraculous qualities which Eastern credulity has gravely ascribed to him. Having no hypothesis to support, and considering it his province rather to narrate events than to speculate upon them, he has confined himself to a simple record of facts: leaving his readers in general to draw their own conclusions.

The conquests of the Arabs, under the once formidable name of Saracens, while promulgating the Koran at the point of the sword,—the vast dominions which they acquired,—and their surprising progress in the cultivation of learning,—are themes which have occupied innumerable pens. To have entered into a full detail of these splendid transactions, or described at length the various dynasties which were planted in the ample regions that lie between the shores of the Atlantic and the frontiers of China, would have been an unwarrantable departure from the plan of the present volumes, by encroaching on the history of countries unconnected with Arabia. In following the march of the victorious Moslem from Spain and Morocco to the wilds of Tartary, the author has endeavoured to imitate the speed of the conquerors; and, in tracing the progress of Mohammedan grandeur, as it shifted from its original seat at Medina to richer lands, and until it crumbled down into a number of independent principalities, he has dwelt no longer on these foreign topics than seemed necessary to preserve the chain of narrative unbroken. In this department of his work he has been enabled, from the valuable labours of Major Price, to rectify some errors, as well as to illustrate some points more fully than has been done by Ockley and Marigny, or even by the Arabian annalists Abulfeda and Elmacin.

When the vanquished nations in Asia, Africa, and Europe, after passing in succession across the scene of action, shook off the yoke of the caliphs, the stream of events which had overspread nearly half the globe necessarily contracts itself within the natural bounds of Arabia. In describing the government, character, manners, and institutions of the present inhabitants, the author need hardly repeat how much he is indebted to the enterprising travellers already mentioned, as well as to others whose names are introduced in the course of the work. This part of the history he considers peculiarly interesting; because, while elucidating the customs and domestic habits of a singular race of people, it developes a state of society with which Europeans have hitherto been very imperfectly acquainted.

The Map which accompanies the work, and the description of the different political divisions of

Arabia in the second chapter, will be found much more full and accurate than any that have yet been laid before the British public. Besides the routes of the pilgrim caravans as laid down by Burckhardt, and the Itinerary of Captain Sadlier, who crossed the desert from the Persian Gulf to the Red Sea, much topographical information respecting the interior has been obtained from the expedition of Mohammed Ali against the Wahabees. The Chart of Nejed, which was constructed to illustrate the campaigns of the Egyptian army, and the treatise of Jomard on the Central Geography of Arabia, appended to Mengin's History, have brought to light much that was entirely new, and corrected various errors with regard to the true position of several places, as well as in the statistics of certain provinces, which our geographers had either left totally blank, or strown with towns and villages on no better authority than the reports of the natives. These improvements and discoveries have been carefully transferred to the prefixed Map; and their value will readily be appreciated when compared with the common geographical delineations of the Arabian deserts.

The chapter on Natural History, it may be proper to remark, is merely intended as a popular view of the subject. To the merit of a scientific treatise it does not aspire; the purpose of the author being chiefly to illustrate the manners and customs of the people, while describing the physical structure and natural productions of the country.

Edinburgh, *July*, 1833.

CONTENTS

OF

THE FIRST VOLUME.

CHAPTER I.

INTRODUCTORY VIEW OF ARABIAN HISTORY

CHAPTER II.

DESCRIPTION OF ARABIA.

CHAPTER III.

PRIMITIVE INHABITANTS OF ARABIA.

CHAPTER IV.

ANCIENT KINGS OF ARABIA.

CHAPTER V

CHARACTER, MANNERS, AND CUSTOMS OF THE ANCIENT ARABS

CHAPTER VI.

LIFE OF MOHAMMED.

CHAPTER VII.

THE KORAN.

ENGRAVINGS IN VOL. I.

ARABIA;

ANCIENT AND MODERN.

CHAPTER I.

INTRODUCTORY VIEW OF ARABIAN HISTORY.

Arabia—Peculiarities in its History and Manners—Interesting Aspect of the Country—Its Connexion with many Scenes and Events in Holy Writ—Distinguished as the Birthplace of Mohammed—Rapid and extensive Conquests of the Saracens—Instability and Downfall of their Empire—Their singular Passion for Learning—Munificent Endowment of Schools—Causes why their History has been little studied in Europe—Ignorance of their Language and Literature—Religious Prejudices against their Character—Efforts of Scholars and Literary Associations to illustrate Arabian History—Valuable Discoveries of recent Travellers—Unexplored Tracts in the Central Deserts—Prospects of further Discoveries—Increased Facilities for Modern Research—Reflections on the Preceding Survey.

To those who delight to study man in his pastoral simplicity, to moralize on the destiny of nations or the rise and fall of empires, the history of Arabia cannot fail to be attractive. From time immemorial it has been celebrated foı its precious productions, and distinguished as the home of liberty and independence; the only land in all antiquity that never bowed to the yoke of a foreign conqueror. It continues to be inhabited at this day by a race coeval with the first ages of mankind. Their manners still present that mixture of rude freedom and patriarchal simplicity which we find in the infancy of society before art had taught men to restrain the sentiments of nature or disguise the original features of their character. This extraordinary people have not only preserved inviolate the dominion of their deserts and their pastures; they have also, with a singular tenacity, retained from age to age, and in spite of changes and revolutions, the vices and virtues, the habits and customs of their ancestors, without borrowing improvement from the progress of knowledge or their intercourse with other nations.

The physical aspect of this country is not less interesting than the peculiar character of its inhabitants. Covered with vast plains of barren sands, intersected by ranges of mountains and fertile valleys, it unites the extremes of sterility and abundance, and enjoys a variety of climate that gives it at once all the advantages of the torrid and the temperate regions. There smiling plenty is often found imbosomed in the midst of desolation; and the indigenous productions of climes the most distant and different from each other flourish there in equal perfection. These grand and distinctive features of Arabia have suffered little alteration from the lapse of time, or the contingencies of human events. Centuries have passed over it without leaving any changes but those produced by the hand of nature. It presents few of those moral vestiges of servile degradation, or those melancholy ruins of departed splendour, that abound in almost every other kingdom in the world. It has, indeed, remains of cities and towns that tell us of a wealth and a population long since vanished; but it has no monuments of art to be compared with the stupendous and imperishable architecture of Egypt, or the classic temples of Greece and Italy.

It possesses, however, scenery of another description, and associations that speak more home to our hearts and our affections than the proudest monuments reared by human labour: with its deserts and mountains are entwined some of our most ancient and hallowed recollections, as places memorable in Scripture history, and consecrated in the eyes of all civilized nations by having witnessed the visible descent of the Divine Being, and some of the sublimest manifestations of his power. It was in Arabia that those wonderful transactions took place which immediately followed the exode of the Israelites from Egypt; its waters were miraculously divided for their passage, it was through its rocky defiles and barren sands that they journeyed for eight-and-thirty years, doing penance for their murmurings and rebellion, before they could be admitted into the Promised Land. The fleets of Solomon and Hiram frequented its seas, and traded in its markets; importing thence the gold and the ivory of which we read in the chronicles of the times. Its traffic and its merchandise are renowned both in sacred and profane history; and for many ages it continued to be the only connecting link of commercial intercourse between the nations of the East and the West.

The inspired writers have borrowed from its manners and its productions some of their finest allusions and most striking descriptions. They make frequent reference to the tabernacles of Edom, the flocks of Kedar and Nebaioth, the incense of Sheba, and the treasures of Ophir. The bride in the Song of Songs draws her imagery from an Arab tent, when she speaks of her beauty as "dark but comely," and compares her tresses to the fine hair of the mountain-goat. The terrible denunciations of the prophets, and the sublime compositions of the Hebrew poets, are greatly indebted to the same source for many of their most pointed and impressive similitudes. Isaiah, in predicting the downfall of Babylon, "the glory of kingdoms, the beauty of the Chaldees' excellency," heightens the picture of its utter desolation by a single allusion to the habits of this pastoral people: "Neither shall the Arabian pitch tent there; neither shall the shepherds make their fold there."—Chap. xiii. 20. No one, in short, can be ignorant how many valuable illustrations the inspired penmen have derived from Arabia, and how much light may be thrown on different parts of the Sacred Scriptures by an attentive observation of the customs and institutions of this and the neighbouring countries. "In order," says the learned Michaelis, "to understand properly the writings of the Old Testament, it is absolutely necessary to have an acquaintance with the natural history as well as the manners of the East; for in that volume we find nearly three hundred names of vegetables: there are many also drawn from the animal kingdom, and a great number which designate precious stones."* The remark of this great biblical scholar is corroborated by an observation of the intelligent Burckhardt to the same effect; "that the sacred historian of the children of Israel will never be thoroughly understood so long as we are not minutely acquainted with every thing relating to the Arabian Bedouins and the countries in which they move and pasture."†

But the principal feature in the history of Arabia consists in its being the birthplace of that extraordinary personage whose artful fanaticism gave a new religion to his country, and produced a revolution which, in its effects on the destinies of mankind, finds no parallel in any age ancient or modern. Prior to the era of their Prophet, the Arabs seem not to have ventured much beyond their own deserts, nor to have made any figure as a great or enter-

* Preface to his Questions addressed to the Danish Travellers.

† Life and Mem. of J. L. Burckhardt, p. lxxxiv.

prising people. Then, however, their history acquired a new interest, and their natural energies took a new direction. Esteemed hitherto of no repute by foreigners, except for their wealth; and separated from all the world, not more by their peculiar mode of life than by a necessity consequent on their situation; we find them suddenly emerging from their national insignificance, and assuming all at once the lofty character of apostles and legislators to the rest of mankind. The sword or the Koran was the terrible alternative they offered to the choice of their enemies. Doubly stimulated, by a thirst for conquest and a zeal for making proselytes, they performed exploits which made their name the terror of the whole earth for many centuries, and have rendered it famous to all posterity.

Nothing, indeed, in the political annals of mankind presents a more extraordinary spectacle than the sudden and overwhelming revolution which, about the middle of the seventh century, sprang up in this obscure corner of the East. Originating in the bold but impious pretensions of one man, who had the art to concentrate the scattered and impetuous energies of his country into the channel of his own ambition, it spread with amazing celerity, and in less than a hundred years covered an extent of territory greater than was ever owned by Rome in the Augustan age of her power. All that we read of the fabled monarchies of Assyria and Babylon, of the boasted expeditions of Cyrus and Alexander, or the vast regions overrun by the Mogul and Tartar hordes, will bear no comparison to the dominion of Mohammed; for it embraced them all. Reaching from the Pillars of Hercules on the one hand to the confines of China on the other, it comprehended, during a certain period, three-fourths of Asia, the whole of Northern Africa, and a considerable portion of Europe.

It is true that the stability of this colossal power

did not equal its greatness. Religious disputes, and the jarring interests of families or individuals who claimed an hereditary title to the succession, gave rise to discords and revolts that soon broke down this huge pontifical monarchy into a variety of separate and independent principalities. At a later epoch, too, foreign invasion completed that overthrow which intestine divisions had begun. The quarrels of rival caliphs were succeeded by wars and revolutions not less sanguinary than had marked the rise and establishment of their power. Greeks, Turks, and Tartars, numerous as the locusts from their own deserts, poured in their wild and undisciplined swarms on all sides of the Moslem dominions, and in process of time won back the extensive territories which a warlike superstition had wrested from them. New states and kingdoms sprang from this imperial wreck, and gradually settled themselves over the fair and ample regions which the Saracen conquests had embraced. The power and magnificence of the caliphs shrank back into the same obscurity from which they had risen. But while their temporal dominion was reduced to its ancient limits within the seas of Arabia, the faith and the fame of Mohammed were left to enjoy all the ascendency which his first triumphs had gained. The victorious nations who threw off the yoke of his feeble successors retained all their veneration for his religion, and willingly rendered him allegiance as their spiritual master; and at the present day his creed reigns throughout the East with nearly as absolute and undisputed authority over the hearts and consciences of men as in the proudest era of Saracen despotism.

Short as was this career of this military pageant, which achieved such vast and extraordinary changes in the moral and political state of a large portion of the world, it is replete with events interesting to the statesman and the philosopher; unfolding a series of characters and incidents that will both engage and reward our curiosity. The victories, revolutions, and capricious vicissitudes of human fortune that pass by in rapid succession are without example in any nation of ancient or modern Europe. The catalogue of the leading personages, the caliphs and conquerors that figured on this remarkable theatre, presents some strange contrasts to the ordinary history of successful adventurers, and the distribution of earthly grandeur. Among other nations, heroes and legislators generally require a process of training, and it is only by slow and persevering degrees that the usurper ascends the pinnacle of his ambition. Here we have the rare spectacle of slaves mounted on thrones; lawless bandits becoming the dispensers of justice and protection; illiterate shepherds and merchants suddenly transformed into the commanders of armies, or vested with the solemn functions of kings and pontiffs. Yet, singular as it may appear, not a few of them were distinguished for civil and military talents; others have gained a lasting celebrity by their patronage and love of science; and some of them shed a lustre on the diadem, by the exercise of those peaceful and princely virtues which have procured for the rulers of other countries the venerable title of fathers of their people.

It was in the courts of Bagdad and Cufa, of Damascus and Cordova, that learning found a hospitable asylum, when a succession of barbarous inroads had nearly quenched the last rays of Greek and Roman literature, and scarcely left a single monument of art or genius in Europe. Nothing, except their own victories, is more surprising than the progress which this acute and ingenious people made in the cultivation of every department of human knowledge. From a state of ignorance and barbarism, in which they had been plunged for centuries, they emerged with a lustre not more remarkable for its brilliancy than for the gigantic height to which it rose. Nor can we account, except from the strength and versatility of their mental capacities, for this sudden blaze of genius which burst forth in every corner of their empire, and spread its influence as far as their arms extended. Many of the caliphs were protectors of learning. They lived surrounded with poets and orators, and assembled in their palaces men of the most distinguished acquirements from every quarter of the world. The name of Haroun-al-Raschid, the hero of the Thousand and one Nights, stands associated with those enchanting fictions which have made Bagdad a fairy land, and will continue to diffuse a charm until taste and imagination shall become extinct.

Under his successors, learning of all kinds was cultivated and propagated with equal zeal. In every town, from the banks of the Tigris to the Atlantic, schools and colleges were established. The sun of science and philosophy diffused its humanizing influence over the fierce spirits and savage manners of Africa. A chain of academies stretched along the whole Mediterranean shore; and in the cities of Cairo, Fez, and Morocco, the most magnificent buildings were appropriated to public instruction.

Spain was one of the most celebrated seats of Arabian learning. A vast number of eminent names in poetry, medicine, mathematics, and every department of study, adorned its annals even in the dreary night of the twelfth century. In its schools and libraries the sacred fires of oriental knowledge continued to burn with more than their ancient splendour, when the rest of the world was sunk in Gothic ignorance.

This ascendency of the Arabs in the empire of letters had followed exactly the progress of their arms; but, like their other conquests, it rested on an insecure basis, and proved quite as transitory in its duration It did not vanish, however, without leaving many important benefits to succeeding generations. Not only were the literary treasures of antiquity preserved, and transfused into the copious language of Arabia; there were also imported new and useful discoveries in arts and manufactures, which had long enjoyed in the East a perfection then unknown to other nations. Every branch of study,—history, geography, criticism, the belles lettres, the natural and moral sciences,—received valuable accessions from the enterprise and enthusiasm of the Arabs. The number of comforts and even luxuries for which we are indebted to them is prodigious; but as they were introduced gradually and at a remote period, we continue to enjoy the benefit of them unconsciously, and without recognising their authors. Many arts and inventions, which have ministered to human happiness, and wrought a total change in the system of human knowledge, were first communicated and taught to us by these Eastern invaders; among whom they were well known long before any indications of them had reached the darkened shores of Europe.

In casting our eye with a rapid glance over the great and diversified events which offer themselves to our contemplation in the study of Arabian history, it is matter of surprise that, until of late, the attention of Europeans should have been so little excited towards this primitive and extraordinary race; and that so much yet remains to be explored of the country they inhabit, and the institutions by which they are governed. Distance, no doubt, and a rare intercourse, have contributed to blunt that curiosity which once looked with astonishment on the trophies of their valour and their learning. New alliances have been formed with the Eastern world; and considerations of trade,—which has long since abandoned its ancient route,—have drawn away our sympathies from this romantic land.

The interest we take in remote or ancient nations

depends greatly on the degree of intimacy with which their memory or their achievements are associated with our present habits. The preference is naturally given to those countries whose language and manners are familiar to us, and incorporated as it were with our ideas and institutions. It is from this circumstance that the Greeks and Romans have engrossed so large a portion of our study. Their authors are the guides and text books of our education. From them, it may be said, we have borrowed the rudiments of our literature, our philosophy, our laws, and our civilization.

With the Arabs the case stands very differently; and they might, not without reason, complain of an apathy which has allowed a veil of ignorance and prejudice so long to rest on their country and their true character. Their language forms no elementary part of our studies, and is seldom approached, except from motives of professional necessity, or a taste which is more admired for its singularity than its usefulness. No Eastern tongue, except the Chinese, is so little cultivated or understood in Europe as the Arabic, notwithstanding the efforts of translators and grammarians,—of Schultens, Reiske, Golius, Erpenius, De Sacy, D'Herbelot, Casiri, Asseman, Pocock, Gagnier, Ockley, Sale, Jones, Richardson, Price, and various others, who have contributed by their learned labours to clear a passage for future adventurers into this vast treasury of oriental knowledge.

There are certain impressions, too, with which Europeans have been wont to associate the character of this ancient and celebrated people, which have had an unfavourable effect, and attached a sort of stigma to their very name. We are accustomed to regard them in the light of robbers and pirates merely, "*whose hand is against every man*;" whose primeval destiny was a sentence of implacable and ceaseless hostility with their neighbours: and on whose fortunes there has rested, since the days of Abraham, the doom of a rejected and expatriated race. Religious prejudices have combined with the predictions of Scripture and the physical impediments of their country, to widen this gulf of ignorance between the Christian and the Mohammedan world.

When the bloody wars of the Koran had ceased, and the chivalry of France and Spain had delivered Europe from the terror of the Saracen arms, a new race of combatants appeared on the field. A host of doctors and schoolmen crowded the theological arena, and fought against the profaners of tombs and the oppressors of pilgrims with all the characteristic bitterness of their sect. Long after the defeat of the Crusaders had left these infidels in the undisputed possession of their mosques and their sepulchres, the ban of the church stood recorded against them;* and the profoundest fathers in Christendom exercised their vast erudition in detecting and refuting the "lies, perjuries, and blasphemies" of the Arabian Prophet; and predicting the final triumph of the Cross over the profane symbol of the Moslem heresy.† Kings and emperors entered the ranks of

* The Greek Church carried its excommunication so high as to pronounce, in their catechisms, *anathema* against the deity worshipped by the Mohammedans, whom they represented as a *solid* and *spherical* being; for so they translated the word *Al-Semed*, applied in the Koran to the deity, and which signifies also *eternal*. The emperor Andronicus ordered this anathema to be erased from the ritual of the Greek Church, on account of the offence it gave to the Saracens who had embraced Christianity. But the Christian doctors opposed it most strenuously. After long and bitter disputes on the subject, the bishops assembled in council, and consented, though with the utmost reluctance, to transfer the imprecation in their catechisms from the "God of Mohammed" to the impostor himself, his doctrine, and his sect.—*Annales Nicetæ*, lib. vii. p. 113. *Reland de Relig. Mohammed*, lib. ii. sect. 3.

† The titles of most of these erudite works vouch sufficiently for the spirit of their contents. The Cardinal Nicolas de Cusa,

controversy, and bound themselves by solemn vows for their extirpation. Military orders were expressly instituted for the same object, who made it a work of charity and mercy to harass and destroy them from the face of the earth. Towards the close of the middle ages their name was detested over all Europe, where they were known only as barbarians and freebooters, the burners of libraries, the Huns and Goths of the East, and the enemies of the Catholic faith.

The improvements and discoveries of navigation, which threw a new light on the most distant quarters of the globe, tended for a while rather to prolong than to dispel the shades of prejudice that had settled down on Arabia. When the naval enterprise of the

about the year 1460, produced his "Cribratio Alcorani," or Sifting of the Koran, as an antidote against the false religion that had made such disastrous inroads into the papal dominions. In 1487, Joannes Andreas, a converted Mussulman, wrote in Spanish his "Confusion of the Mohammedan Sect," in refutation of the creed he had forsaken. So early as 1210 Friar Richard had gone to Bagdad in order to confute the Mohammedans out of their own books: and, on his return, he published his "Confutatio Legis Saracenicæ." The Jesuit missionary Xavier, at the command of the Mogul emperor, wrote a defençe of the Gospel against the Saracens, called "A Looking-glass for showing the Truth." A learned Persian wrote an answer to it, entitled the "Brusher of the Looking-glass," in which he exposed the errors and superstitions of the Church of Rome, so as to alarm the Propagandists, who employed a Franciscan friar to refute it; and, in 1628, he published his "Clearing up of the Looking-glass," in reply to the Persian Brusher. Among other defenders of Christianity who directed their polemical fury against the Saracens, may be mentioned Alphonso de Spina, who wrote the "Fortalitium Fidei," or Fortress of the Faith; Raymond Martin, with his "Pugio Fidei," or Dagger of the Faith; and Pelagius, with his "Collyrium contra Hereticos," or Eye Salve against the Heretics. Some of these productions are poor enough, and give a curious picture of the extravagant fancies of the writers. Among the early specimens of English typography, we find a book printed by Daye, "Agaynst Perjurius Murderying Mahomet."—*Watt's Biblioth. Brit. Prideaux's Life of Mahomet.*

Portuguese had, in 1497, opened up a new passage to India by the Cape of Good Hope, the genius of ancient commerce abandoned the shores of the Red Sea; and its altered course dissolved, in a great measure, the only remaining bond of connexion which had hitherto attracted the cupidity of strangers towards the fertile regions of gold and frankincense.

In process of time, however, these barriers, which ignorance and bigotry had reared, were gradually broken down. Literary men, both here and on the Continent, began to study the works of the Arabian authors; and to unlock, by means of their translations and commentaries, the sealed treasures of oriental learning. Pocock, at Oxford, spent more than sixty years in this laudable toil.* Various foreign scholars trod in his footsteps with equal erudition; and to them the world is indebted as the first restorers of Arabian literature. Since their time numerous societies have been instituted for the purpose of collecting and giving to the public such information respecting the East as may lie scattered among the hoarded stores of modern libraries; or can be procured by the research of individuals who have visited or resided in the countries whose history, manners, and productions it is their object to illustrate. Several Arabian works have recently appeared under the auspices of the College of Fort William in India, and of the Oriental Translation Committee established in London. Within the last two or three years, a periodical work in their own language was projected in the capital of France, for the benefit of the Arabs, in order to

* The elder Pocock (Edward) is a name of which English literature may well be proud. He studied in the East, and on his return was dismissed by his master with this compliment,—"Go thy way; wheresoever thou goest, thou knowest more Arabic than the Mufti of Aleppo."—*Ockley's Hist. Pref.* vol. ii p. 28.

make them acquainted with the literature and politics of Europe; and to rekindle, as it were, the torch of knowledge in those neglected deserts where the arts and sciences drew their first breath.* The Committee of Public Instruction at Calcutta are daily occupied in translating into the Eastern tongues the most esteemed writings of modern authors, both English and Scottish.

Much was also done for the elucidation of oriental history by the liberal and enlightened spirit of British commerce during the last century. At an early period the merchants trading to the Levant formed themselves into a company, which was acknowledged and protected by the government. Their powers were great, and their intercourse extensive, including the states of Barbary and all the shores of the Mediterranean eastward of Sicily. The benefits conferred by this association in the way of research and illustration were immense; and it is to their consuls, chaplains, and agents, many of whom were individuals of very distinguished talents, that we owe our best and earliest knowledge of the countries connected with their traffic.†

It is almost superfluous to remark, that several valuable additions to our stock of information have been made by the adventures of modern travellers. The journals of Niebuhr and Burckhardt alone, though there are many other useful narratives, have done more to illustrate the geography, manners, and customs of Arabia, than all that has been written since the revival of letters.‡ Its botany and zoology

* Asiatic Journal, vols. viii. and xxii.

† Among the agents of the Levant Company who have conferred obligations of this kind, we find the names of R. Pococke, Ricaut, Maundrell, Smith, Shaw, the two Russels, Porter, Dallaway, &c.—all intelligent travellers.

‡ Niebuhr, the Danish traveller, visited Arabia in 1763 Burckhardt in 1810–17; and Badhia, a Spanish Mussulman, under the name of Ali Bey, in 1807.

have been investigated, its ancient ruins examined, and its present condition described; so far at least as came within the range of their own observation, or could be gathered from their converse with the natives. The recent war with the Wahabees opened up new channels of discovery, by affording to several Europeans attached to the Egyptian army an opportunity of penetrating farther into the Arabian deserts than had been deemed prudent or practicable by solitary travellers. The survey which these expeditions enabled them to make has illustrated many curious and doubtful points, and brought us acquainted with extensive pastoral tracts little known, and almost totally unexplored by strangers; for the knowledge of which we had to depend on Arabian authors, often very incorrect; or on the Greeks and Romans, who described these immeasurable wilds chiefly from fanciful and exaggerated reports.*

The early pilgrims from Europe and Africa, who annually visited the grand temple at Mecca, had neither means nor leisure for observation; they performed their devotions, and, however enlightened as to their spiritual prospects, returned at least entirely ignorant of the country. The Crusaders met the Saracens only as enemies in the field of battle, where they found them enthusiastically brave, often generous in the hour of victory, and always faithful to the laws of honour and hospitality. The French and Portuguese, by their several expeditions to the Red Sea and the Persian Gulf in the 16th and 17th centuries, made geographers acquainted with little more than the coasts, on some parts of which they had effected temporary settlements.

* The pen of Herodotus once celebrated this country, and might have supplied many defects in its early history. But the work has long since perished in the wreck of classical learning; and the loss is the more to be regretted when we reflect on what he has done for the antiquities of Egypt.

It is but of late that feelings of ancient prejudice have ceased to haunt the imagination of European travellers. The Chevalier D'Arvieux, French consul in Syria from 1682 to 1688, visited the camp of an Arabian emir, and mentions his surprise at the polite civility of these Eastern savages, whom the people of France had been taught to consider as having nothing human about them but the shape; while the prince of the desert and his courtiers were equally astonished to find that the Franks, whose names they used to frighten their children, were not cannibals, nor quite so barbarous as had been represented.* Niebuhr and Burckhardt, who have earned such an honourable distinction in this interesting field of research, concur in their admission that the pictures drawn of Arab ferocity, and the dangers to be apprehended from it, have been greatly overcharged; and that travellers, when they meet with incivilities or injuries, have usually themselves to blame, either by affecting an ostentation of wealth and consequence, which acts as an incitement to plunder; or by expecting such luxuries and conveniences as are utterly incompatible with the simple habits and resources of the country.

But with all this laudable and successful enterprise, the labour of Arabian discovery is far from being completed. There is yet scope for exertion. The prying eye of observation, which has made important disclosures in several of its unfrequented provinces, has left various central districts nearly unexplored, and as little known as they were in the days of Alexander the Great. There is, however, small reason to doubt but that this obscurity will gradually disappear. Though there are few allurements to tempt the literary adventurer, compared with the dangers he runs, and the necessary qualifications of mind and body; still there is enough of

* Voyages de l'Arabie Deserte, Pub. par M. De la Roqu

scientific chivalry to carry forward, if not to complete, the discoveries that have been so auspiciously begun. Persevering research has lifted up, or rather for ever removed, the veil that so long hid the mystic writings of ancient Egypt. A new interest has dawned on the ruins of her tombs and her temples. Those monuments of nameless kings and gods, instead of remaining the objects of barren admiration or blind enthusiasm, have been restored to history and chronology; and those primitive fountains of learning to which Greece herself owed the rudiments of her knowledge have again become the oracles of wisdom to the Western world. It required indefatigable patience, and a multitude of fatal attempts, to trace the hidden sources of the Nile and the Niger. Similar perseverance may dispel the ignorance that covers the interior of Arabia; and though it cannot make the wilderness and the solitary place to bud and blossom as the rose, it may restore to geography much that the wreck and the negligence of a thousand years have buried in oblivion.

The task can be approached now with many facilities that have sprung from the improvement of art, or grown up with the liberal and enlightened progress of society. The history of the terrible Saracens can be discussed with more candour and freedom than in the reigns of Omar and Saladin. A recital of their cruelties will find, in our day, no Peter the Hermit to make them a theme of declamation for rousing the pious zeal of kings and emperors; or for letting loose a second time upon Asia the undisciplined fury of a superstitious multitude. Even their religion may be spoken of without reviving those apprehensions which alarmed the piety of Prideaux and the bigotry of Maracci. A fair representation of its doctrines will hardly expose us to the spiritual attacks of those *daggers* and *fortresses of the faith* with which the Christian doctors of yore were wont to assail their antagonists. The tolerant spirit of our age has effaced the prejudices against a difference of belief which ignorance and fanaticism had created. The sword—the grand argument of the Moslem religion—has yielded to the force of reason; and our manners and habits of thinking have triumphed in their turn over the relentless soldiers of Mohammed.

The subject, embracing such a variety of events, is necessarily extensive. We shall endeavour to collect within a moderate compass every thing which, from its novelty or importance, deserves to be recorded; and if our limits do not permit us to gather all the flowers that lie scattered over the surface of this pleasant landscape, we hope at least to be able to produce some of those treasures of solid information which, like gold mixed with sand or buried in mountains, have been alloyed with Eastern fable, or concealed from the general reader by being wrapped up in dead or foreign languages. We have thought it essential to our plan to give some account of the dark and traditionary epoch that preceded the time of Mohammed, in order to preserve a connected chain of narrative with the more brilliant and authentic events that followed it. Besides, it is impossible, without such aid, to understand either the literature or the religion of the Arabs. Their tales and their poetry abound with images, the origin of which must be sought beyond the memory of written records. Even the Koran itself has perpetual allusions, not only to the facts, but to the fables and traditions which the stream of antiquity had mingled and carried down in its course. What we shall say of the government, customs, and institutions of the ancient Arabs shall be restricted to what is absolutely necessary to a right understanding of their civil and political history after they had risen to the dignity of a warlike and powerful empire.

CHAPTER II.

DESCRIPTION OF ARABIA.

Name—Boundaries—General Features—Ancient Geographical Divisions—Arabia Petræa—Deserta—Felix—Modern Divisions—Hejaz—Tehama—Yemen—Hadramaut—Oman—Lasha or El-Hassa—Hejed—Peninsula of Sinai—Ancient Bed of the Jordan—Mounts Sinai and Horeb—View from their Top—Various Opinions as to their Identity—Climate of Arabia—Heat—Rains—Rivers—Winds—The Simoom—Arabian Seas—Persian Gulf—Red Sea—Coral Banks—Passage of the Israelites—Dangerous Navigation—Steam Communication with India.

ARABIA is generally allowed to have derived its name from a Hebrew word, denoting a wilderness or land of deserts and plains. Various other derivations have been assigned; and learned etymologists are divided in opinion, whether the term be expressive of a mixed, a mercantile, or a western people. The Arabs themselves trace it to one of their ancestors, whom they call Yarab, a son of Joktan, who is said to have been one of the earliest settlers in that country; but as Yarab does not occur among the thirteen sons of that patriarch mentioned in Scripture (Gen. x. 26–29), this inference may be considered as purely traditional. The name of Arabah is repeatedly applied to the western wilderness by Moses, who describes it with a minuteness not to be mistaken, as situated "over-against the Red Sea, between Paran and Tophel: and by the way of Elath and Ezion-gaber."* A small tract in the

* The word translated *plain* (Deut. i. 1, and ii. 8) is in the original *Arabah*, by the Red Sea, &c. "Arabia non ab Arabo,

ancient Idumæa still retains the original appellation, and as these territories belonged to the wandering Ishmaelites, the name would gradually be extended as they spread their conquests over the rest of the country. By this name it is recorded in the writings of the Jewish historians and the later prophets, who speak of the kings of Arabah, of its traffic, and the different tribes by which it was inhabited. (Josh. xv. 52, 61; 1 Kings x. 15; Jer. xxv. 24.)

At this remote period were these western regions distinguished from the more fertile and populous plains towards Chaldea, which went by the name of Kedem or the East,—a distinction as old as the days of Abraham and Job. This simple practice of deriving names from territorial residence is entirely in accordance with the notions that regulated the primitive divisions of the earth, when mankind had no other geography than such as respected their own local situation, or the relative position of the heavens. The ancient Greeks called Italy Hesperia, or the Land of the West; the Italians bestowed the same epithet on Spain; and the name was at length transferred to those fabulous gardens, which gradually retired before the dawn of knowledge into the Elysian solitudes of the Atlantic Ocean. Similar ideas prevail in the East at the present day. Syria is uniformly called Sham,—the country to the left, or the north; while the south is termed Yemen, or the country to the right. The Turks and Persians call the whole peninsula Arabistan; the natives themselves call it Jezirat el Arab (the peninsula of the Arabs); and it is remarkable as one of the few countries among the kingdoms of antiquity which, amid the changes and revolutions of 3000 years, still

Apollinis et Babyloniæ filio, et Latinorum plerique asserunt, sed ab Araba, plaga non longe a Medina sita, nomen sortita est."—Gab. Sionita, de Mor. et Nat. Orient. p. 7. Univ. Hist. vol. xviii. chap. 21. Gagnier, ad Abulfed. Geog. Diatrib. de Arab. nom. sect. 1.

retains the precise appellation which it bore within a few centuries of the deluge.

This vast tract lies between latitude 12° 45′—34½° north, and longitude 31°-60½ east from Greenwich. Its form is that of an irregular triangle, surrounded on three sides by water. Eastward, its limit is the Persian Gulf and the Euphrates; on the south lies the Indian Ocean; on the west the Red Sea divides it from Egypt, Nubia, and Abyssinia. The northern frontier is not so well defined, and has been subject to considerable variations. The ancients restricted it to an imaginary line, stretching between the extreme points of the Arabian and Persian Gulfs. The rest they attached partly to Egypt and partly to Syria. But the conquests and settlements of the Arabs have long extended their territory beyond this ideal boundary. On the authority of Burckhardt, the northern frontier may be taken as a line running from Suez across the isthmus of that name to the Mediterranean, near El Arish, passing along the borders of Palestine and the Dead Sea, and thence winding through the Syrian desert by Palmyra until it reaches the Euphrates above Anah, the course of which river it follows till joined by the Tigris; at which point their united streams take the name of Shut el Arab, or boundary of Arabia. Part of the northern frontier lies now within the pashalic of Damascus, which extends as far south as Tor Hesma, a high mountain, one day's journey from Akaba.

The Greek and Roman geographers prescribed a limit somewhat different. Xenophon carries it beyond the Euphrates, including the greater part of Mesopotamia, or the Arabian Irak; Ptolemy bounds it by the Chaldean mountains on this side the river, and northward by the city of Thapsacus, near Racca. The same is adopted with little variation by Diodorus and Strabo. Abulfeda, an Arabian geographer who wrote about the beginning of the fouteenth century, extending the northern boundary somewhat higher than Burckhardt, places it at Beles, nearly in the latitude of Aleppo. The length of this extensive region, from Anah to the Strait of Bab el Mandeb, is reckoned about 1480 miles; and its middle breadth, from Suez to Bussora, above 900. On the south, it presents a base of 1200 miles washed by the Indian Ocean.

In its general features Arabia may be described as an elevated table-land, sloping gently towards the Persian Gulf. The whole of the southern coast is a wall of naked rocks, as dismal and barren as can well be conceived. Here and there they imbosom a low sandy beach, but they are entirely destitute of soil or herbage, offering to the eye of the mariner a striking picture of ruin and desolation. The mountains, brown and bare, rise in several ranges, one behind another, to the height of 1000 or 1500 feet. Such is the impenetrable rampart, dark, waste, and wild, with which nature has guarded the fabled land of "Araby the Blest." On every other side this peninsula is encircled with a belt of flat, dry, sandy ground; that on the north is composed of the Hauran (Auranitis) or Syrian Desert; that on the east, of the level shores of the Persian Gulf.

The interior of the country is chiefly burning deserts, lying under a sky almost perpetually without clouds, and stretching into immense and boundless plains, where the eye meets nothing but the uniform horizon of a wild and dreary waste. Over the face of this vast solitude the sand sweeps along in dry billows, or is whirled into hills and columns, having the appearance of waterspouts, and towering to a prodigious height. When the winds leave them at rest, they resemble the ocean; and their level expanse, at a small distance, is sometimes mistaken by the thirsty traveller for a collection of waters. This deception recedes as he journeys on, keeping always in advance; while the intermediate space glows like a furnace, occasioned by the quivering undulating motion of that quick succession of vapours and exhalations which are extracted by the solar rays. Every object is magnified to the eye, insomuch that a shrub has the appearance of a tree, and a flock of birds might be mistaken for a caravan of camels. The most singular quality of this vapour (Sirab), or mirage as it is termed, is its power of reflection,—objects are seen as from the surface of a lake, and their figure is sometimes changed into the most fantastic shapes.

These naked deserts are encircled, or sometimes intersected, by barren mountains, which run in almost continuous ridges, and in different directions, from the borders of Palestine to the shores of the Indian Ocean. Their summits tower up into rugged and insulated peaks, but their flinty bosoms supply no humidity to nourish the soil; they concentrate no clouds to screen the parched earth from the withering influence of a tropical sky. The refreshment of cooling breezes, periodically enjoyed in other sultry climates, is here unknown. The air is dry and suffocating. Hot and pestilential winds frequently diffuse their noxious breath, alike fatal to animal and vegetable life. The steppes of Russia and the wilds of Tartary are decked by the hand of nature with lofty trees and luxuriant herbage; but in the Arabian deserts vegetation is nearly extinct. The sandy plains give birth to a straggling and hardy brushwood; while the tamarisk and the acacia strike their roots into the clefts of the rocks, and draw a precarious nourishment from the nightly dews. An inspired pen has truly described this steril country as "a land of deserts, and of pits; a land of drought, and of the shadow of death; a land that no man passed through."—(Jer. ii. 6.)

This general aspect of desolation is occasionally relieved by verdant spots, or valleys with little streamlets, lying among the hills, and formed by the alluvial depositions of the winter showers. These wadys (the oases of the Greeks), which appear like islands in the trackless ocean, are both fertile and pleasant. Their rich verdure and groves of date-trees supply food and pasture for the roving colonies of the wilderness. There are, besides, various wells or watering stations, partly natural, partly artificial, on the routes which traverse the deserts in several directions; serving as points of intercourse between distant parts of the country. Without these reservoirs the greater portion of Arabia must have remained unpeopled, and for ever impervious to man. Their brackish waters afford refreshment for the weary pilgrims, and enable small hordes of settlers to cultivate patches of ground, on which scanty crops or a few common vegetables are reared These tanks or reservoirs are often built of stone, and form the usual resting-places of travellers and caravans: the water is raised in leathern buckets by means of an iron chain passed over a pulley, and drawn by cows or oxen. It is sold to strangers on their journey, and is often transported to a considerable distance on the backs of camels. Among the Arabs, water constitutes a great part of their wealth. It is the most valuable property in districts of fifty or a hundred miles round. The possession of a spring has occasioned hot disputes, and even been the cause of civil wars. We read of Abraham rebuking Abimelech because his servants had violently taken possession of a well; and of the strife between the herdsmen of Gerar and those of Isaac. It is also mentioned as an instance of intolerable tyranny in one of the ancient Arab kings, that he would suffer no camels but his own to be watered at the same place. There are entire districts, however, where this luxury, as it may well be called, is unknown. The great southern desert, which extends from six to seven hundred miles in length, and as much in breadth, does not possess a single fountain of water.

From the singular local situation of Arabia, the inequalities in the nature of its soil and climate may readily be inferred. Though the central portion consists of arid and burning wastes, the aspect of the country in other parts is materially different. In the south, more especially where the land is broken into hills and valleys, there are tracts of remarkable fertility, which enjoy a succession of almost perpetual verdure. So short is the interval between the decay and reproduction of vegetable life, that the change is scarcely perceptible. Though nature perhaps nowhere realizes those splendid landscapes which borrowed their colouring from fancy rather than truth, and converted this happy region, in the minds of foreigners at least, into an earthly paradise, still the picture offers an agreeable contrast to the contemplation of dreary sands and desolate rocks. The air is more temperate, while the rains and dews descend more copiously. The hills are wooded to the tops, or covered with a rich alpine turf. From their sides fall perennial streams, sometimes in beautiful cascades, which run a course of considerable extent among cultivated fields or luxuriant gardens. Fruits of all kinds are delicious and abundant. The fertility of the earth at once invites and rewards the industry of the husbandman; and nature, by lavishing her choicest favours here, seems to have compensated for her want of hospitality everywhere else. What Waller says in his Summer Island is true of these delightful regions:—

"The gentle spring, that but salutes us here,
Inhabits there, and courts them all the year"

Such is a general outline of what may be termed the physiology of this celebrated peninsula.

The political divisions of Arabia are differently represented by different writers. The knowledge which the ancients had of the country was imperfect, and little reliance can be placed on their descriptions. They scattered mountains, cities, and rivers over its surface at random; chiefly, it would seem, to fill up a void in their maps, and to preserve a kind of symmetry or analogy between this and the other portions of the earth with which they were better acquainted. Even D'Anville's accustomed accuracy is here at fault; and travellers have frequently borrowed from others what they had not the means to verify by actual survey. Within the ast quarter of a century the torch of war has thrown a new light on many parts of the desert which might have still remained unexplored, had not these regions become the theatre of hostilities with a foreign enemy. The Turkish geographers divide Arabia into twelve provinces, while others limit them to two. The division most familiar to us is that introduced by Ptolemy, viz. the Stony, the Desert, and the Happy Arabia; a distinction which is applicable to the general features of the country rather than descriptive of separate provinces. The Greeks, it is well known, took great liberties with countries of which they had little acquaintance; but here they ought not to be accused of imposing arbitrary names, since they merely translated words which have a similar import in the original, and had been used by the natives themselves.

Petræa, or the Stony Arabia, occupied the mountainous tract between Palestine and the Red Sea It was the country allotted to Esau or Edom, from whom it took the name of Idumea. It was the land of the Amalekites, Midianites, Hittites, Hivites, Kedarenes, Hagarenes, Nabathæans, and other tribes descended from Abraham, so often mentioned in the wars of the Jews, under Moses, Joshua, and David. To this wild but interesting region belongs a reverence which no other portion of the earth, Judæa excepted, can claim. It was the theatre of many awful and extraordinary events recorded in Jewish history. The sacred eminence of Sinai, on whose cloudy summit the Deity made his pavilion of darkness when he first issued a system of written laws to the human race—Horeb, with its burning bush, and its caves that gave shelter to Elijah when he fled from the persecution of Jezebel—the pastoral solitudes where the Jewish deliverer, then an exile from Egypt, kept the flocks of Jethro, the priest of Midian—Shur and Parah, with the bitter wells of Marah, and the smitten rock that yielded water—the land of Uz, the scene of the wealth and the woes of Job, of the trial of his patience and the triumph of his piety—are all comprehended within the geography of Petræa.

Arabia Deserta extended north and east as far as the Euphrates. It was separated from Petræa by the ridge of Mount Seir, and understood to comprehend the great central wilderness; but its limits were vague and obscure.

Arabia Felix embraced the celebrated Region of Incense on the coast of the Indian Ocean. The ancients have dwelt with all the extravagance of romance on the costliness of its productions, and the wealth and number of its inhabitants. Marcian informs us, that in his time it contained fifty-four provinces, one hundred and sixty-four towns and villages, fifteen ridges of hills, four considerable rivers, five bays, two seacoasts, with thirty-five adjacent islands.* Strabo states that it was divided into five kingdoms, and that its chief cities abounded in temples and palaces.† The principal nations men-

* Arrian, Marcian, Agatharcides, Dionysius, Periegetes, &c. may be consulted in Hudson's Geograph. Minor. See also Dr Vincent's learned Disquisitions on the Periplus of the Erythræan Sea.

† Strabo, Geog. lib. xvi. Pliny (Nat. Hist. lib. vi.) says that

tioned by Pliny, Ptolemy, and the ancient geograph ers as settled here were the Minæi, Sabæi, Atramitæ, Catabeni, Maronitæ, Homeritæ, Sapphoritæ, Omanitæ, and a variety of others whose names and localities it is almost impossible to identify with any modern tribes or provinces in Arabia. Dionysius Periegetes, who wrote a description of the world in Greek verse, has celebrated this thuriferous region of the "lovely Arabia" (Ἀῤῥαβιης ἐρατεινης), where the fields were decked with undying verdure, and the atmosphere loaded with spicy odours. Of the forty-two cities mentioned by Abulfeda, and the six hundred by Ptolemy, the most ancient and populous were situated in Arabia Felix. The imagination of the Greeks, easily set on fire, pictured in golden dreams the ideal wealth of this Arabian Tempe; but it is evident that foreign nations knew little more of the country than its name, and that it abounded in gold, gems, spices, perfumes, with other natural rarities, the value of which was greatly enhanced by their own ignorance and cupidity.

The three divisions introduced by Ptolemy, and still adopted in modern geography, are unknown to the Arabs themselves; who, like the Egyptians, Turks, and Persians, would find it difficult to recognise their own names in the tongues of Europe. From time immemorial, this peninsula has been parcelled out into various independent territories; but as it never, properly speaking, has formed one kingdom, the number or limits of these provinces have not been very exactly defined. Regions are sometimes divided from each other by a solitary shrub, and the extent of hereditary property determined by the distance at which a dog can be heard to bark.

Charmæi, a town of the Minæi, was fourteen miles in compass; that a city of the Agarturi was twenty miles about; and that Sabotale, the capital of the Atramitæ, had sixty temples within its walls.

The grand geographical divisions laid down by Niebuhr, and more minutely described by subsequent travellers, are eight; though considerable tracts of country are not strictly or politically included in these territorial departments.

Hejaz, the Holy Land of the Moslems, lies on the middle coast of the Red Sea; its chief city, Mecca (the Macoraba of the Greeks), being the capital of the Mohammedan religion. It is a barren district, consisting of sandy plains towards the shore, and rocky hills in the interior; and so destitute of provisions as to depend even for the necessaries of life on the supplies of other countries. Among its fertile spots is Wady Fatima, which is well watered, and produces grain and vegetables. Safra abounds in date-trees. Taïf, seventy-two miles from Mecca, is celebrated for its gardens; and the neighbourhood of Medina has cultivated fields. The towns on the coast are Jidda and Yembo; the former being considered the port of Mecca, from which it is distant about fifty-five miles; and the latter that of Medina. Hejaz is bounded eastward by a lofty range of mountains, which, near Taïf, take the name of Gebel Kora. The scenery there is occasionally beautiful and picturesque; the small rivulets that descend from the rocks afford nourishment to the plains below, which are clothed with verdure and shady trees. The vicinity of Mecca is bleak and bare; for several miles it is surrounded with thousands of hills, all nearly of one height; their dark and naked peaks rise one behind another, appearing at a distance like cocks of hay. The most celebrated of these are Safa, Arafat, and Meroua, which have always been connected with the religious rites of the Mohammedan pilgrimage. The whole of this territory may be considered as almost an absolute desert; D'Anville assigns it an extent of coast of about 750 miles.

Tehama is the flat sandy belt that extends along

the Red Sea, nearly from Akaba to Aden; though it is chiefly restricted to the maritime plains southward of Hejaz. It stretches backwards to the mountains, varying in breadth from thirty to eighty miles. It bears every mark of having been anciently a part of the bed of the sea, from which it has gradually emerged. The soil is interspersed with marine fossils and other exuviæ, and contains large strata of salt, which in some places shoot up into hills. As the sea continues to recede, the Tehama extends its limits in proportion. The coral banks gradually increasing, and the intermediate space being filled with accumulating sands, new ground is thus formed, and annexed to the continent. But this conquest over the watery element is of little advantage to man, as the land is altogether unsusceptible of cultivation. Tehama is by some included in the two adjoining provinces, and therefore not reckoned a separate territory. Between it and Hejaz lies the large district of Abu-Arish.

Yemen corresponds nearly to the ancient Arabia Felix, and still comprehends the finest and most fertile portion of the peninsula. Hali, on the Red Sea, divides it from Hejaz. It presents considerable diversity of soil and climate; towards the coasts it is scorched and barren, but the interior is a highland country, full of precipitous yet fertile hills, and enjoying a healthy and temperate air. Its extent is reckoned at 20,000 square miles. It is parcelled out, in unequal portions, into a great number of petty sovereignties, of which Niebuhr has enumerated fourteen; the principal being Aden, Kaukeban, Khaulan (supposed to be the Havilah of Scripture), Sahan, Nejeran, Kahtan, Heschid-u-Bekil, Jôf or Mareb, and Jafa. Some of these are subdivided into three or four minor states, which are often ruled by independent princes. Sanaa, Mocha, Loheia, Taas, Hodeida, Zebid and Damar are the chief towns; all of which have been visited by European travellers.

HADRAMAUT (the Hazarmaveth of Scripture, Gen. x. 26) lies eastward of Yemen, which it greatly resembles in its soil and surface. The hills are extremely fertile, and intersected with well-watered vales. Its towns were more celebrated, and perhaps better known, in the time of Strabo and Ptolemy, than they are at present. It was famous in the days of Augustus for the bravery of its inhabitants, and still more for being the country whose mountains produced frankincense. This extensive province was included in Arabia Felix; its harbours are Keshin, Merbat, Dafar, and Hasec, in the great gulf of Kuria Muria, which is surrounded with isles. Doan and Aidan, are towns in the interior; but of their situation we have no accurate knowledge. Shibam, which seems to be the Saba of the ancients, is described as the seat of a powerful prince, eight days' journey from Sanaa, and ten from Mareb. Niebuhr heard more than twenty cities mentioned, of which he could learn nothing beyond the names; but these, he observes, bore a striking resemblance to those recorded by the most ancient historians—a circumstance which renders it probable that this region has undergone little change since the remotest ages. The mountainous tracts called Seger (or Sahar) and Mahrah are comprehended in this province.*

OMAN occupies the eastern angle of the peninsula between Hadramaut and the mouth of the Persian Gulf. It is filled with mountains, which almost everywhere extend to the sea. Muscat (the Mosca of Arrian) is the capital, and the best known to Europeans of all the Arabian cities on the Indian Ocean. The whole coast in this quarter exhibits the same bleak and steril aspect already noticed;

* Niebuhr, Descript. Arab. tom. iii. p. 160–254.

presenting, in many places, a precipitous rocky wall towards the sea, alongside of which ships might float in safety. The name of Oman implies a land of peace or security, as contrasted with the unciv ilized and inhospitable countries by which it is bounded. It is said to extend from near Cape Mussendom, on the north-west, to the island of Mazeira, south of Ras el Hud, which is literally the "Land's End" of Arabia. Its breadth is reckoned six days' journey towards the south-west. Throughout this space are scattered towns, villages, and hamlets, in great abundance. The mountains are in general rugged and bare, but very lofty. Hence the dews and clouds which they arrest give a mild and agreeable temperature to the air; while the showers, washing down the decomposed surface of the earth, add to the soil of the valleys, and also occasion rills and torrents to fertilize them. The people, from their situation on the Persian Gulf, have been celebrated from the earliest dawn of commerce. Their character, and the different positions on their coast, are described with considerable minuteness in the ancient periplus of the Erythræan Sea. Northward of Cape Mussendom lie the territories of Seher and Julfar; neither of which are properly included in the province of Oman.

LAHSA, or more properly EL HASSA, stretches along the Persian Gulf as far as the mouth of the Euphrates. This district is also denominated Hajar, and sometimes Bahrein; but the latter appellation is now restricted to the islands of that name in the adjacent gulf. The coast is flat and dreary, diversified here and there with groves of palm-trees, that indicate the site of towns or villages, of which the number contained in the province is reckoned twenty. Its breadth inland is only fifty or sixty miles. It is celebrated for its numerous wells, some of which are covered over with vaulted roofs, supported by tall white marble columns, seen at a great distance. Rich clover pastures abound, which supply food to a fine breed of Arabian horses. The rivulets are fringed with lilies and privets; but the country suffers fearful encroachments from the drifting sands, by which whole cantons are sometimes invaded. The principal town is El Hassa, a place of some note as one of the strongholds of the Arabs in their late wars with the pashas of Egypt and Bagdad. Another considerable town is Koneit or Graen, said to contain 10,000 inhabitants. El Katif is supposed to be the ancient Gherra, which was a famous *entrepôt* for the spices and perfumes of the south. Taroot, a small town to the east of El Katif, has excellent vineyards, which are sometimes flooded by the tides. It is here we must place the Regio Macina of the Greek geographers, where the vines, raised in earthen pots or baskets of rushes, were subject to be carried off by the waters of the sea.* The inland boundaries of these maritime provinces are far from being accurately defined. Nature, however, has set limits to them in the immense central desert of Southern Arabia, called Akhaf, which extends from the mountains behind Tehama to the frontiers of Oman, and is perhaps one of the most dreary regions on the face of the earth. The Arabs give it the name of Roba el Khali, or the "Empty Abode."† This vast expanse of sand contains nearly 50,000 square miles, and has no supply of water except from the clouds. The skirts of this frightful wilderness produce herbage when refreshed by the winter rains; but its depths have never been explored. One single station, the Wady Jebrin, on the route to Hadramaut, diversifies this solitary tract; it has wells and date-

* Strabo, lib. xvi. p. 528.

† Burckhardt's Travels in Arabia, Appendix, No. iv. According to the tradition of the Arabs, this desolate region was once a terrestrial paradise, where dwelt a race of giants, who, for their impiety, were swallowed up by a deluge of sand.

trees, but its noxious climate renders it unfit for habitation.

NEJED is the largest province in Arabia, occupying the great central desert (the Arabia Deserta), from Hejaz on the west to the narrow strip of El Hassa on the east. Until within the last twenty-five years it was nearly a blank in the maps of Europe; or filled with names at random, according to the recital of travellers who had never visited it. Its breadth, according to Captain Sadlier, who traversed the whole peninsula from sea to sea in 1819, cannot be less than 750 miles. According to Jomard, the distance between the two gulfs, in a straight line from El Katif to Yembo, may be estimated at 270 leagues; the extent of the Nejed, from north to south, he reckons at 260.* The surface is diversified with mountains and plains; but it is by no means that barren and desolate region which it has been hitherto represented. On the north, from the Hauran to the banks of the Euphrates, the whole tract is one immense level, called El Hamad, without the slightest elevation, and showing no trace of town or village; but affording vigorous growth to a few thorny shrubs, by which the traveller's eye is sometimes relieved. Xenophon's description of these regions, which were successively trodden by the armies of Cyrus and Julian, is as applicable at the present day as it was nearly 2300 years ago. "The country," says the Greek historian, "was a plain throughout, as flat as the sea, and full of wormwood; and if any other shrubs or reeds grew there, they had all an aromatic smell; but no trees could be seen. Bustards, ostriches, antelopes, and wild asses appeared to be the only inhabitants."† Southward

* Memoirs of the Lit. Soc. of Bombay, tom. iii. Notice Géographique, par M. Jomard, Mengin, vol. ii. Appendix.

† 'Εν τούτῳ δὲ τῷ τόπῳ ἦν μὲν ἡ γῆ πεδίον, ἅπαν ὁμαλὸν ὥσπερ θάλαττα, ἀψινθίου δὲ πλῆρες· εἰ δέ τι καὶ ἄλλο ἐνῆν ὕλης ἢ καλάμου, ἅπαντα ἦσαν εὐώδη, ὥσπερ ἀρώματα· δένδρον δ' οὐδὲν ἐνῆν.—*Xenoph. Anab.* lib. i. cap.

s the desert of Akhaf; so that Nejed is surrounded on every side by immense sandy plains. Many parts in the interior are well watered, and celebrated for their excellent pastures. There are also many remains of ancient buildings, of very massive structure and large dimensions, but in a state of complete ruin. Innumerable wadys are imbosomed in the hills, which abound with fruits and grain of the finest quality. Yet the occasional want of rain causes frequent scarcities, which happen every three or four years, and are generally accompanied with epidemical disease. The country of Nejed is subdivided, according to Mengin, into twelve or fourteen smaller provinces, among which some include El Hassa.

Ared,	Haryk,	Dowaser,
Woshem,	Kharaj,	Soubeah,
Soudeir,	Gebel,	Aflaj,
Kasym,	Shahran,	Beishe.

Ared contains the capital city, Deraiah; but it is not so populous as Beishe, nor so fertile as Kasym, which is the richest province in Nejed. Kharaj is the same as the district of Yemama, and its chief town is still called by that name. The most remarkable mountains in Nejed are Shammar, Khora, Salma, Shahak, and Toweik. Shammar stands near the great route from Damascus, and is, properly speaking, a district by itself. It rises behind a sandy desert, and is covered with forests and villages. In height and extent it resembles Mount Lebanon, and is seen by the pilgrims at a great distance. Above a hundred towns and villages have been enumerated in this extensive region; but this amount is probably much underrated, as there are large tracts still unexplored by Europeans. What

22.—*Gibb. Rom. Hist.*, chap. xxiv.—British officers still find the same species of game that amused the soldiers of Cyrus.—*Keppel's Narrative*, vol. i. chap. v.; *Sadlier's Itiner.*

has been done, however, towards the elucidation of this province, by the industry of Burckhardt, De Sacy, Jomard, and Mengin, is enough to enlighten the ignorance of Pinkerton, who confounded Yemama with Yemen, and thought that Nejed (which signifies the Highlands, as distinguished from Tehama, or the Lowlands) was so named from a mountain, and might be regarded as unknown to European geography.

The PENINSULA OF SINAI may be considered a province of Arabia, though not reckoned one of its political divisions. No part of that country has been so minutely explored or so elaborately described as this interesting region. Its general aspect is singularly wild, and well merited the name of Petræa. A recent traveller describes it as "a sea of desolation. It would seem," says he, "as if Arabia Petræa had once been an ocean of lava, and that while its waves were literally running mountains high, it was commanded suddenly to stand still."* The whole of this wilderness is a collection of naked rocks and craggy precipices, interspersed with narrow defiles and sandy valleys, which are seldom refreshed with rain or adorned with vegetation. Fountains or springs of water are extremely rare; and those that do exist are either brackish or sulphurous, but of a wholesome quality. Some of the plains are covered with loose flints and pebbles. Few shrubs or plants are to be met with, and those that are found are indebted to the clefts of some barren rock, or a thin mixture of clay in the soil, for their support. The ridge of mountains called Seir and Hor in Scripture stretches from the borders of the Dead Sea towards Ailah. On the western side runs the long valley Wady Ghor and Wady Arabah,

* Sir Frederick Henniker's Notes on Egypt, &c. See also Burckhardt's Travels in Syria, Pococke's Description of the East Shaw's Travels, Fazakerly's Journey to Sinai, in Walpole's Travels, &c

from three to twelve miles in breadth; this, without doubt, formed the ancient bed of the Jordan, which must have emptied itself into the Arabian Gulf before the terrible overthrow of the Cities of the Plain arrested its waters in the pitchy lake of Sodom and Gomorrah. On the eastern side is Wady Mousa, the site of the ancient Petra, which gave its name to the district. Mount Hor is the highest of the chain, on the summit of which is pointed out the tomb of Aaron (Num. xx. 28), still held in great veneration by the Arabs.

Near the centre of the peninsula stands the group of the Sinai Mountains, the upper region of which forms an irregular circle of thirty or forty miles in diameter. It is difficult to imagine a scene more desolate and terrific than that which is discovered from the top of Sinai. Nothing is to be seen but huge peaks and crags of naked granite, composing, as far as the eye can reach, a wilderness of steep and shaggy rocks and valleys destitute of verdure. Yet in the highest parts of these regions water is to be found, and fertile spots which produce fruit-trees. This sacred mountain consists of two elevations, Gebel Mousa and Gebel Katerin, which are generally identified with Sinai and Horeb. Both terminate in a sharp peak, the planes of which do not exceed fifty or sixty paces in circumference. The latter is the higher of the two, and its summit commands a very extensive prospect of the adjacent country, the two arms of the Red Sea, a part of Egypt, and northward to within a few days' journey of Jerusalem. There is some doubt, however, as to which of these the appellation is to be assigned; some conjecturing that the lower eminence is Sinai, while others are of a contrary belief.

This confusion has arisen from the circumstance that the several names have been indiscriminately applied to this mountain. The manner in which Moses uses them, as convertible terms, has led to the supposition that the two must be twin summits of the same hill; an opinion for which there does not appear to be any solid foundation, since Horeb may be interpreted, and seems to have been used, as the name of a rocky district or desert country, rather than the proper name of any particular eminence. The language of Scripture would lead us to suppose that Sinai was a detached mountain in the midst of a plain, and that the Israelites encamped around it. Its immediate vicinity afforded pasturage for their cattle, otherwise it would have been impossible for them to have remained so long in that quarter; and its name suggests that it abounded in some species of acacia. Josephus describes it as an extremely pleasant place, and the discontented Israelites sojourned here twelve months without murmuring. These incidents certainly do not well correspond with the steril neighbourhood of Gebel Mousa. "It is not easy to comprehend," says Niebuhr, "how such a multitude as accompanied Moses out of Egypt could encamp in these narrow gullies, and frightful and precipitous rocks; but perhaps there are plains that we know not of on the other side of the mountain." There are valleys, however, at no great distance, where their flocks might find pasture; and Shaw speaks of "a beautiful plain more than a league in breadth, and three in length, closed to the southward by some of the lower eminences of Sinai. In this direction, likewise (he adds), the higher parts of it make such encroachments upon the plain that they divide it into two, each of them capacious enough to receive the whole encampment of the Israelites." Some travellers have observed, that were this the real Sinai, it would be found to exhibit traces of the stupendous phenomena which attended the manifestation of the Divine presence, in the visible symbols of fire, and earthquake, and apparent volcanic eruption. Burckhardt, however, could not detect the slightest vestige of these supernatural appearances, though there are islands in the Red Sea which retain marks of volcanic action. But objections such as these are entitled to little weight. We do not read of any actual discharge from the mountain. It is described, indeed, as having "quaked greatly," as having "burned with fire," and emitted smoke "like a furnace:" but these appearances were not the effect of any natural convulsion; they were the awful emblems which the Deity chose to make the harbingers of his presence,—the cloudy robes of his divine majesty. To look, therefore, for the ordinary results of such phenomena in the site of this wonderful and miraculous transaction were as reasonable as to expect to find the nightly pillar that enlightened the Hebrew camp, or the fountains which followed them on their march through the wilderness.

Another still more elevated summit westward, called Mount Serbal, has been considered as having rival pretensions to the honours claimed for the Mount of Moses; but these do not seem to be better founded. It exhibits no volcanic appearances; and its five peaks, according to the general theory on the subject, militate against the idea that it is the Horeb and Sinai of Scripture. That it was first selected as the representative of Sinai was probably owing to its great elevation; but the hypothesis is altogether gratuitous which considers that sacred mountain as pre-eminently high. Notwithstanding the many theories and conjectures of travellers, the probability remains stronger in favour of Gebel Katerin and Gebel Mousa than of any other; however difficult it may be to reconcile their scenery or position with the several events recorded in Holy Writ. There is yet room for future investigation; and it is possible that, by examining the different localities with more attention to the Sacred Record than to the legends of monks and Bedouins, further light may yet be thrown on this interesting geographical problem.

Little dependence can be placed on local tradition. Burckhardt expresses his disappointment at being able to trace so very few of the ancient Hebrew names of the Old Testament in the modern geography of the peninsula. With the exception of Sinai and a few others, the appellatives are all of Arabic derivation; and the incongruous association of Moses and St. Catherine is a proof how little reliance is to be placed upon them. Sinai is two or three times mentioned in the Koran; but in neither instance is there any reference to its relative locality. "Call to mind," says Mohammed (chap. ii.), "when we accepted your covenant, and lifted up the mountain of Sinai over you;" alluding to a ridiculous legend, that when the Israelites refused to receive the law of Moses, God tore up the mountain by the roots, and shook it over their heads, to terrify them into compliance.

The three highest eminences in this peninsula are St. Catherine, Serbal, and Shomar. To the two latter has also been attributed the distinction of having witnessed the promulgation of the decalogue. Burckhardt ascended Mount Serbal, though he had no means of ascertaining its elevation. The upper region is described as almost perpendicular: it is approached from below over sharp rocks, without any path, and climbed by means of steps in several parts, cut through the rock with great labour, or regularly formed with large loose stones. The summit of the eastern peak consists of one immense mass of granite, the smoothness of which is broken only by a few partial fissures, presenting an appearance not unlike the ice-covered summits of the Alps. When seen from the bottom, it looks as sharp as a needle, but it terminates in a platform of about 130 feet in circumference. The surface of every block presented inscriptions written in characters of a foot in length, the greater part of which are illegible. These facts, together with the road leading up to the top, afford strong grounds to presume that this singular eminence was an ancient place of devotion. "From these circumstances," says our intelligent traveller, "I am persuaded that Mount Serbal was at one period the chief place of pilgrimage in the peninsula; and that it was thus considered as the mountain where Moses received the Tables of the Law; though I am equally convinced, from a perusal of the Scriptures, that the Israelites encamped in the Upper Sinaï, and that either Gebel Mousa or Mount St. Catherine is the real Horeb."

Om Shomar lies more towards the south, and nearer the point where the two gulfs separate. It rises in a mass of almost perpendicular cliffs, in a country the aspect of which is that of the most savage wildness. "The devastations of torrents are everywhere visible,—the sides of the mountain being rent by them in numberless directions. The surface of the naked rocks is blackened by the sun; all vegetation is dried and withered, and the whole scene presents nothing but utter desolation and hopeless barrenness." The highest peak tapers to a point, and appears to be inaccessible; but at 200 feet below, a beautiful view opens on the Gulf of Suez. Whether Sinai may be identified with this eminence or not, it is probably the same range of mountains; and the idea is not extravagant that would consider the bold promontory of Ras Mohammed the seaward front of the Mount of God. This rugged and lofty chain is visible from both arms of the gulf; and perhaps, as a lively traveller has remarked, the fisherman in his bark must have heard the thunder and seen the cloudy pavilion when the God of Israel spoke to his chosen people.*

* Scenes and Impressions in Egypt, p. 63.

The whole of this rugged tract is intersected with innumerable wadys, some of which are nearly as barren as the rocks; while others nourish shrubs, fruit-trees, and occasionally a most luxuriant vegetation. The valley of the mountainous range, called El Tyh, which forms the northern boundary of the Sinai group, affords excellent pasturage and fine springs, though not in great numbers. Wady Leja, near Gebel Mousa, is represented by Burckhardt as most delightful. It is small; but so brilliant was the verdure, and so aromatic the perfume of the orange-trees that grew in an orchard, that he fancied himself transported from the cliffs of the wilderness to the delicious groves of Antioch. It is supposed to be the valley of Rephidim, and opens into an extensive plain towards the north-east.

Wady Sheik, and its continuation Wady Feiran, which separates Mount Serbal from the Upper Sinai, is considered the finest valley in this part of the Arabian peninsula. From the higher extremity an uninterrupted succession of gardens and date-plantations extends downwards for several miles; and almost every one of these has a well, by means of which the grounds are irrigated during the whole year. The waters collected from the lateral ravines empty themselves through Feiran into the Gulf of Suez. Wady Kyd, between Shomar and the Gulf of Akaba, is a very romantic spot, and one of the most famous date-valleys in the district. It is traversed by a small rivulet, overshadowed by trees, with fine verdure on its banks. The rocks that overhang it almost meet, and give to the whole the appearance of a grotto. Similar descriptions might be extended to numerous other valleys; but enough has been said to convey a tolerable idea of the nature of this remarkable peninsula.

A region so extensive as Arabia, varying in elevation, in climate, and soil, must naturally be subject to considerable irregularities of temperature, as well as of natural productions. While the inhabitants of the plains and valleys suffer from heat, and enjoy perpetual abundance, those on the mountains are obliged to wrap themselves in sheepskins, and subsist by plunder. In the desert the thermometer is generally above 100° during the night, at 108° in the morning, and in the course of the day it rises to 110°, and sometimes higher, even in the coolest and best shaded parts. All travellers who have visited the coasts of the Red Sea appear to have been oppressed by the extraordinary heat, and to have considered the temperature of other tropical countries as moderate in comparison. Burckhardt remarks that the climate of Mecca is sultry and unwholesome; the rocks that enclose its narrow valley interrupt the northern breezes, and reflect the rays of the sun with redoubled intensity. The air at Medina is much colder in winter; but in summer it is said that the heat is greater here than in any other part of Hejaz. At Mocha, it averages from 90° to 95° in July; owing to its vicinity to the arid sands of Africa, over which the south-east wind blows for so long a continuance as not to be cooled in its short passage across the strait.* In Muscat the thermometer varies from 92° to 102° during the day, and the heat of the night is felt to be almost equally oppressive and unfavourable to European constitutions. Among the mountains of Petræa the diversity is much greater; while, in the upper regions, the maximum in May was 75°; in the lower country, and particularly on the seashore, it stood from 102° to 105°, and sometimes at 110°. In the desert, near the Euphrates, Griffith observed that the variation in the thermometer, from two to four in the day and the same hours in the morning, was frequently sixteen degrees; and that, during the prevalence of the land-winds, it rose to 132° under

* Valentia, vol. ii. chap. 8.

the tent, and 156° when exposed to the sun's rays.* The highlands on the coast, and in some parts of the interior, enjoy a more temperate atmosphere. Near Sanaa, about 200 miles inland from Mocha, Niebuhr was informed that ice had been seen. Storms of hail are not uncommon at Taïf; and snow sometimes falls on the hills near Medina. In winter the whole of the Upper Sinai is often covered with snow; many of the passes are choked up, so that the mountains of Moses and St. Catherine are inaccessible. Mr. Fazakerley, who ascended them in the month of February, found it very deep; though he fared better than Pietro della Valle, who went up in a violent snow-storm, and gives a lamentable account of his adventures on that occasion. For this peculiarity of climate Arabia is partly indebted to its position, hemmed in between the continents of Asia and Africa, and effectually debarred by the latter from the influence of the south-west monsoon, which blows during summer on the coasts of India, and ushers in the periodical rains.

One great characteristic of this vast continental desert is aridity. Whole years occasionally pass away without rain; the drought is consequently extreme, and destructive of all vegetation. All the highland tracts, and the different ridges which shoot forth into the interior, by attracting clouds and vapours, enjoy the advantage of frequent and copious showers. Those rains occur at different times of the year, according to the position of the mountains. On the western declivity of Yemen, and along the shores of the Red Sea, they commence in June and terminate in September. This district is also refreshed by a spring rain; while on the eastern declivity of the same mountains the wet season is between the middle of November and the middle of February In Hadramaut and Oman, and on the coasts of the

* Travels in Arabia, p. 384.

Persian Gulf, it extends from the middle of February to the middle of April. Thus it would seem that the rains make the tour of the peninsula every season according to the prevalence of the winds. They often fall, however, in storms rather than showers; and, instead of irrigating the ground, are drunk up by the thirsty sands, or collected in sudden pools. In the valleys, near Taif, Burckhardt was overtaken by a tempest of thunder, hail, and rain, which covered Wady Noman three feet deep; innumerable cascades immediately tumbled from the sides of the hills, and the inundation became general, so as to render travelling for a time impossible. The historians of Mecca record various instances in which that town was completely deluged. In 1626 a torrent rushed so rapidly into the plain that five hundred of the inhabitants were drowned; the great mosque was filled; three sides of it were swept away; and every human being within it perished. There appears to be no general or fixed law by which these periodical rains are determined; and it is only the skirts, as it were, of the Arabian peninsula that enjoy this necessary provision of nature for sustaining the productive powers of the earth. The same latitudes in Asia and Africa present the same peculiarities. Persia, except where it is watered by the Euphrates and the Indus, exhibits all that frightful sterility which has been depicted by the historians of Alexander in recording the perils and sufferings of his army while traversing Gedrosia (Mekran). But for the Nile, Egypt were a desert; and if Barbary is more fertile than Sahara, it is because the Atlas range attracts the moisture of the clouds.

Perhaps the most singular feature in the Arabian continent is its entire want of rivers or perennial streams. This deficiency has indeed been generously supplied by the industry of geographers, who have traced winding lines in various directions, terminating, after a long course, on the margin of the ocean. Ptolemy reckoned four rivers in Arabia Felix; Diodorus and Strabo describe several fine streams; and Herodotus speaks of one traversing the desert, at the distance of twelve days' journey. By the times of D'Anville and Niebuhr these had greatly diminished; and modern travellers have discovered that names which have so long flourished as pompous rivers are either quite imaginary, or only temporary currents, which are absorbed in the sand, and never reach the sea except after copious rains. The great Aftan of Ptolemy, on which stood the city of Yemama, and which is still made to roll its tributary waters into the Persian Gulf, is now found to be a very modest brook, nourished by the clouds, and having no existence but during one season of the year.* Those at Aden, Mocha, and other places are of the same description. The Jews and poorer inhabitants erect their huts of wicker-work in the dusty channels. In some of the wadys there are streams of considerable size that run a course of sixty or eighty leagues; but they are generally drunk up in the sandy belt before reaching either gulf. The lakes in the interior, mentioned by the Greek and Turkish geographers, must have been temporary collections of water formed by the rains.

The winds are extremely variable, and their refreshing influence is but partially felt. During summer, the heat in the lower plains on the coast is so steady and equable that the atmosphere remains in a state of repose. No change of temperature takes place to set the air in motion; hence dead calms occur which sometimes continue for sixty days without interruption. The nature of the winds differs according to the point of the compass from which they blow, or the tract over which they respectively

* Notice Géographique sur l'Arabie Centrale; Mengin's Histoire de l'Egypte, vol. ii. p. 551; D'Anville, Géog. Ancienne

pass. On the shores of the Persian Gulf, the south-east wind is accompanied with a degree of moisture which, when the heat is intense, occasions violent perspiration, and on that account is deemed more disagreeable than the north-west, which is more torrid, and heats metals in the shade. Water placed in jars, exposed to the current of this hot wind, is rendered very cool by the effect of the sudden evaporation; but its blasts often suffocate both men and animals. In the lower part of the Red Sea, the winds blow from the same quarter about nine months in the year, or from the end of August till May; but from Cosseir to Suez the opposite monsoon or north wind prevails.

Arabia is frequently visited by the terrible simoom, called by the natives *shamiel*, or the wind of Syria, under whose pestilential influence all nature seems to languish and expire. This current prevails chiefly on the frontiers, and more rarely in the interior. It is in the arid plains about Bussora, Bagdad, Aleppo, and in the environs of Mecca that it is most dreaded, and only during the intense heats of summer. The Arabs, being accustomed to an atmosphere of great purity, are said to perceive its approach by its sulphurous odour, and by an unusual redness in the quarter whence it comes. The sky, at other times serene and cloudless, appears lurid and heavy; the sun loses his splendour, and appears of a violet colour. The air, saturated with particles of the finest sand, becomes thick, fiery, and unfit for respiration. The coldest substances change their natural qualities; marble, iron, and water are hot, and deceive the hand that touches them. Every kind of moisture is absorbed; the skin is parched and shrivelled; paper cracks as if it were in the mouth of an oven. When inhaled by men or animals, the simoom produces a painful feeling, as of suffocation. The lungs are too rarefied for breathing, and the body is consumed by an internal heat, which often terminates in convulsions and death. The carcasses of the dead exhibit symptoms of immediate putrefaction, similar to what is observed to take place on bodies deprived of life by thunder, or the effect of electricity.

When this pestilence visits towns or villages, the inhabitants shut themselves up, the streets are deserted, and the silence of night everywhere reigns. Travellers in the desert sometimes find a crevice in the rocks; but if remote from shelter, they must abide the dreadful consequences. The only means of escaping from these destructive blasts is to lie flat on the ground until they pass over, as they always move at a certain height in the atmosphere. Instinct teaches even animals to bow down their heads, and bury their nostrils in the sand. The danger is most imminent when they blow in squalls, which raise up clouds of sand in such quantities that it becomes impossible to see to the distance of a few yards. In these cases the traveller generally lies down on the lee side of his camel; but as the desert is soon blown up to the level of its body, both are obliged frequently to rise and replace themselves in a new position, in order to avoid being entirely covered. In many instances, however, from weariness, faintness, or sleepiness, occasioned by the great heat, and often from a feeling of despair, both men and animals remain on the ground, and in twenty minutes they are buried under a load of sand. Caravans are sometimes swallowed up; and whole armies have perished miserably in these inhospitable deserts.

Such are the effects of these resistless whirlwinds; but the noxious qualities ascribed to them, though pernicious to health, have certainly been exaggerated by credulous or ill-informed travellers. Their deadly influence seems to arise solely fom heat contracted in passing across burning wastes; hence, when suddenly inhaled, they occasion sickness and suffocation, and even those livid appearances that have been ascribed to atmospheric poison. The simoom usually lasts three days; but if it exceed that time it becomes insupportable. It blows from the east and the north, and is of such excessive aridity that water sprinkled on the ground evaporates in a few minutes. When the wind changes to the south, every thing is in the opposite extreme,—the air is damp, and substances when handled feel clammy and wet to the touch. The predominating winds in the Nejed are the gharbi, or south-west, which is dry, and pernicious to cultivation, and occasionally blows from the same point seven months in succession; the hesiah, or west wind, is of a burning heat, and prevails in June, July, and August. The shamal, or north, is cool and refreshing; the jenoub and sharki (south and east), "the fathers of the rains," are the welcome harbingers of clouds, which soon dissolve in grateful showers.

A description of Arabia necessarily includes the two gulfs that form its eastern and western boundaries. Both of these seas figure in the early annals of oriental commerce; they are filled with sunken rocks, sandbanks, and small islands, which throw impediments in the way of free and safe navigation. Pliny has remarked, that nowhere are the depositions from rivers more perceptible than at the mouth of the Euphrates. He mentions the famous reservoir, which he calls Baramalchum (Bahr el Malec, i. e. the Royal Lake), formed by Nebuchadnezzar, who raised a mound, or wall, to confine the waters at the mouth of the Tigris. The Persian Gulf is included by Nearchus, Arrian, Strabo, and other Greek writers, under the name of the Erythræan Sea,—so denominated, as they allege, from a certain king, Erythrus, who reigned and was buried in one of the islands at its estuary. Ormuz stands associated with the ancient wealth of India; and Tyrus and Aradus are supposed to be the cradle of the Tyrians and Phenicians. The Bahrein group, on the Arabian coast, have always been, and still are, celebrated for their pearl-fishery. In the neighbourhood of these islands fresh springs are found in the middle of the salt water. The Persian coast is safer and more elevated than the Arabian. Near the upper end the gulf is forty leagues in breadth, and about seventy in the middle; but the strait at Cape Mussendom does not exceed fifty-five miles.

The Red Sea occupies a deep rocky cavity, extending about 1160 miles in length, and its mean breadth may be taken at about 120. Strabo has compared its shape to that of a broad river; and, as has already been noticed, it does not receive the waters of a single tributary stream. The name greatly puzzled the ancients, and has occasioned in later times a display of much superfluous learning to determine whether it was derived from the colour of the water, the reflection of the sandbanks and the neighbouring mountains, or the solar rays struggling through a dense atmosphere. These various conjectures are set at rest; both the air and water are unusually clear; the theory of King Erythrus is exploded; and the name is now admitted to be merely a Greek translation of the " Sea of Edom (a Hebrew word denoting Red), so frequently mentioned by the sacred writers. The surface is diversified with a number of islands; some of which, such as Kotembel, and Gebel Tar, near Loheia, exhibit volcanic appearances. The western coast is bold, and has more depth of water than the eastern, where the coral rocks are gradually encroaching on their native element. These reefs are found dispersed over the whole gulf, rising in some places ten fathoms above the water. The bottom is covered with an abundant harvest of this substance, as well as of certain plants; and, if examined in calm weather, it has the appearance of verdant meadows and submarine forests,—phenomena which procured this gulf the appellation of Yam Zuph from the Jews, and Bahr

Souf with the Arabs, signifying the "Sea of Green Weeds." These beautiful productions attracted the admiration of antiquity. Strabo seems to allude to them when he speaks of trees, resembling the laurel and the olive, growing at the bottom and along the eastern coast of the Red Sea, which at ebb tide were left uncovered, though at other times they were wholly under water; a circumstance deemed the more surprising when contrasted with the nakedness of the adjacent shores. Burckhardt remarks that the coral in the inlet of Akaba is red, and that in the Gulf of Suez the white is chiefly to be seen; facts which may reconcile the discordant statements of Bruce, Valentia, Henniker, and other modern travellers.

All who have frequented the Red Sea have observed the luminous appearance or phosphorescence of its waters. "It was beautiful," says a picturesque writer who sailed from Mocha to Cosseir, "to look down into this brightly transparent sea, and mark the coral here in large masses of honeycomb rock, there in light branches of a pale red hue, and the beds of green seaweed, and the golden sand, and the shells, and the fish sporting round the vessel, and making colours of a beauty to the eye which is not their own. Twice or thrice we ran on after dark for an hour or two; and though we were all familiar with the sparkling of the sea round the boat at night, never have I seen it in other waters so superlatively splendid. A rope dipped in it and drawn forth came up as a string of gems; but with a life, and light, and motion, the diamond does not know."* Those sealights have been explained by a diversity of causes; but the singular brilliancy of the Red Sea seems owing to fish-spawn and animalculæ, a conjecture which receives some corroboration from the circumstance that travellers who mention it visited

* Scenes and Impressions, p. 35.

the gulf during the spawning period,—that is, between the latter end of December and the end of February. The coral banks are less numerous in the southern parts. It deserves notice, that Dr. Shaw and Mr. Bruce have stated,—what could only be true, so far as their own experience went,—that they observed no species of weed or flag; and the latter proposes to translate Yam Zuph, "the Sea of Coral,"—a name as appropriate as that of Edom.

Bab el Mandeb, the narrowest part of the gulf, is the strait at its entrance, which is between twelve and fourteen miles across; it is divided by the island of Perim, which stands about three miles from the Arabian shore. Strabo relates, that the Egyptian merchants who had possession of this sea used to draw a chain across to the African side, to prevent the intrusion of foreigners; an assertion which is probably to be classed among the other marvels of the ancients. The high land of Africa and the Peak of Assab are distinctly visible, although the latter is reckoned seventy miles distant from Mocha. This proves that there is a great degree of refraction in the atmosphere. In further confirmation of this fact, Lord Valentia mentions a singular phenomenon which occurred, and which has also been noticed by the ancients. The setting sun had the appearance of a flaming column, having totally lost its usual round form; a splendid testimony in favour of Agatharcides, who also says that it rose like a pillar of fire.* The northern part of the Red Sea separates into the two gulfs of Akaba or Ailah and Suez, called by the Greeks and Romans the Elanitic and Heroopolitan, from the cities that stood at their extremities. The former is dangerous, owing to its shoals and coral rocks; the common opinion that it

* Valentia's Travels, vol. ii. p. 359. "Nec sol ad disci formam se habet, sed crassam refert columnam principio."—*Hudson. Geograph. Minor. Agathar Diodor* lib. iii. cap. 3.

terminates in two points has been corrected by Burckhardt, no such bifurcation being found to exist.

The Gulf of Suez extends about 160 miles in length, and is of safer navigation; its depth varying from nine to fourteen fathoms, with a sandy bottom. On the Elanitic side, the whole coast, from Ras Mohammed to Akaba, consists of a succession of bays bounded by rocky headlands. Here, as in other parts, the shores have undergone a material change. On the Arabian coast the water has retired, so that towns anciently mentioned as seaports are now several leagues inland. The land at Suez presents evidence that the sea had then extended much farther northward—appearances which tend to favour the hypothesis that the Arabian Gulf was at some remote period a strait which united the Indian and Mediterranean Seas; and that the isthmus which now divides them has been subsequently filled up with sand. The tides and medium level of this gulf are subject to great variation from the influence of the periodical winds; so much so, that Niebuhr tells us the point near Suez may be sometimes crossed on foot.

This western arm of the Red Sea has been rendered famous by the miraculous passage of the Israelites. The exact spot where this event occurred, as well as the line of march and different encampments of the chosen race, have become too obscure through time and change to be traced with accuracy. Shaw and Pococke have given routes of their journeyings, probably copied from older maps; but many of their stations must, of necessity, be matter of conjecture. The natives of the coast point out indifferently the valley of Baideah, nearly seventy miles down; the passage from Suez across the narrow arm that runs up to the port; and other points on the shore farther southward, opposite Ayoun Mousa, and the Hammam Faraoun. Niebuhr fixes upon Suez as the spot at which they crossed. The narrow gulf before that town, he observes, appears at first sight to be only the breadth of a river in comparison with the open sea, and therefore too small to have been chosen by the Almighty as the scene of the manifestation of his power. This led him to suppose that the Israelites must have passed at some leagues southward of Suez; an opinion which he changed on measuring the breadth of the gulf at that place, which he found to be 3500 feet, and farther north it was still wider. "If the children of Israel," he continues, "passed the sea at Kolzoum, the miracle would indeed be less than if they crossed it near Baideah. But it is a mistake to suppose that the multitude could cross here without a prodigy; for even in the present day no caravan crosses here in going from Cairo to Mount Sinai, notwithstanding that it would materially shorten the distance. It must naturally have been more difficult to the Israelites thousands of years ago, when the gulf was probably broader and deeper, and extended farther northward." This supposition of the Danish traveller fairly meets the objection, that if the Israelites had crossed at Suez, Pharaoh might easily have doubled the point and overtaken them, without risking the loss of his army by entering the channel in pursuit. When Burckhardt left Suez the tide was at flood, and he was obliged to make the tour of the whole creek, which he says can be forded at low water; but in winter time, and immediately after the rainy season, the circuit is rendered still greater, because the low grounds to the northward, for many miles, are then inundated, and become so swampy that the camels cannot pass them.

The ingenious Dr. Shaw objects to the opinion which fixes the passage opposite Ayoun Mousa, on the ground that the water there must have been too shallow to drown so many Egyptians; and this objection applies still more strongly to the theory of Niebuhr. According to the first of these learned travellers, the Israelites must have crossed lower down, opposite the desert of Shur. Supposing Rameses to have been Cairo, there are two roads, he remarks, by which they might have been conducted to Pihahiroth on the coast; the one through valleys which are bounded on each side by the mountains of the Lower Thebaïs; the other, more to the north ward, having these mountains for several leagues on the right, and the desert on the left, till it turns through a singular ravine in the northernmost range into the valley of Baideah. The latter he presumes to have been the course taken by the Israelites. Succoth, the first station, signifies only "a place of tents;" and Etham, the second station, he considers as probably on the edge of the mountainous district just alluded to. Here the Israelites were ordered to turn from their line of march, and encamp before Pihahiroth, in the mouth of the gullet or defile between Migdol and the sea. This valley he supposes to be identified with that of Baideah, which signifies miraculous, and which also bears the name of Tiah-Beni-Israel, or the Path of the Israelites. Baal-zephon, over-against which they encamped, is alleged to be the mountain still called Gebel Attakah, or the Hill of Deliverance; and at the distance of ten miles from this is the desert of Shur, where the Israelites landed. The gulf in this quarter would be capacious enough to cover a numerous army, and yet might be traversed by the Hebrew fugitives in a night; whereas lower down, from Wady Gharendel to Tor, the channel is from ten to twelve leagues broad, which is too great a distance to have been travelled by a multitude with so many encumbrances as they carried with them. Having once entered this valley, it might well be said that the wilderness had "shut them in," inasmuch as the mountains of Mokattem would deny them a passage to the southward, while those near Suez would be a barrier towards the north.

Burckhardt seems inclined to follow the opinions of Shaw and Niebuhr. Referring to the distance between Ayoun Mousa and the Well of Howara, he conjectures that this is the desert of three days, said to have been crossed by the Israelites immediately after their passing the Red Sea, and at the end of which they arrived at Marah. "In moving with a whole nation, the march (nearly forty miles) may well be supposed to have occupied three days; and the bitter well at Marah, which was sweetened by Moses, corresponds exactly with that of Howara. This is the usual route to Mount Sinai, and was probably, therefore, that which the Israelites took on their escape from Egypt; provided it be admitted that they crossed the gulf near Suez, as Niebuhr with good reason conjectures. There is no other road of three days' march on the way from Suez towards Sinai, nor is there any other well absolutely bitter on the whole of this coast, as far as Ras Mohammed. The complaints of the bitterness of the water by the children of Israel, who had been accustomed to the sweet water of the Nile, are such as may be daily heard from the Egyptian servants and peasants who travel in Arabia."

With respect to the means employed by Moses to sweeten the waters, Burckhardt frequently inquired among the Arabs in different parts, whether they possessed any means of effecting such a change, by throwing wood into it, or by any other process; but he never could learn that such an art was known. Forskal, who travelled with Niebuhr as botanist, indicates a plant having this property, which is said to be known in the East Indies; and Burckhardt suggests the red berry of the gharkad, a shrub which grows in the neighbourhood, and which he thought might perhaps effect this change, in the same manner as is done by the juice of pomegranate grains. If Howara is the Marah of Exodus (chap. xv. 23), then Wady Gharendel is probably Elim with its twelve springs and seventy palm-trees. But, as we have already observed, it is vain to reason from modern appearances. The retirement of the sea, and the increase of coral shoals, has so much altered the gulf, that no decisive arguments can be built on the present shallowness of the waters, or breadth of the channel. We know that in former times ships entered the harbour of Kolzoum (the Clysma or Arsinoë of the Greeks), but in consequence of the retreat of the waters that place was deserted. Suez, which was not in existence towards the end of the fifteenth century, rose on its ruins. Besides, those who endeavour to account for this phenomenon by natural causes forget that the transaction was miraculous. Even if we suppose that the agency of the tides was employed by Providence in favouring the passage of the Israelites, the east wind, which, blowing all night, divided the waters of the gulf in the middle, laying the channel bare, as between two walls, was clearly supernatural, since the monsoon there blows constantly from the north and the south. And as this unprecedented ebb of the waters must have been supernatural, not less so was the sudden reflux by which the Egyptians were completely overwhelmed.

Contrary to the generally received opinion on this subject, Lord Valentia has started a theory, that the Israelites must have crossed to the northward of Suez; as the presumption is that the marshes, which extend for about twenty-five miles in that direction, were then overflowed with water. This supposition contains nothing inconsistent with Scripture history or with natural appearances; and it removes a difficulty which Dr. Shaw could not reconcile, except by alleging that Josephus had been guilty of making "too hasty statements," in causing the children of Jacob, encumbered as they were with their families, cattle, baggage, and kneading troughs, to perform a march of ninety Roman miles in three days.

The remembrance of this memorable transaction is preserved in the local traditions of the inhabitants. The Wells of Moses (Ayoun Mousa), and the Baths of Pharaoh (Hammam Faraoun), are associated with the names of the Jewish deliverer and the Egyptian monarch; and the superstitious Arabs call the gulf the Bahr of Kolzoum, or Sea of Destruction, in whose roaring waters they still pretend to hear the cries and wailings of the ghosts of the drowned Egyptians.*

Of the navigation of the Arabian seas, the ancients uniformly spoke with awe and apprehension, as everywhere full of peril and difficulty. Arrian, Agatharcides, Strabo, and Abulfeda unite in drawing the same terrific picture of tempests, whirlpools, and sunk mountains, with which these inhospitable waters were infested. The storms dashed their ships on the rocks, and the rocks cut their cables; while the inhabitants were more terrible than either, for they plundered and ate, or made slaves of all who escaped the wrecks and the waves. The first navigators never ventured to encounter these complicated dangers until they had instituted solemn festivals, or performed sacrifice to Neptune; and those who had the fortune to return in safety were regarded as prodigies, and adorned with garlands

* Diodorus seems distinctly to allude to the passage of the Israelites: "It has been an ancient report among the Ichthyophagi, continued down to them from their forefathers, that by a mighty reflux of the waters, which happened in former days, the whole gulf became dry land, and appeared green all over, the water overflowing the opposite shore; and that, all the ground being thus left bare to the very lowest bottom of the gulf, the sea, by an extraordinary high tide, returned again into its ancient channel." (Lib. ii. cap. 3.) It is not unimportant to find a heathen writer unconsciously bearing testimony to the truth of Scripture history.

and crowns of gold. Nearchus, who sailed in the year 326 B. C., by order of Alexander, from the Indus up the Persian Gulf, set out with more pomp of preparation, and met with more hardships and adventures, than Columbus did in circumnavigating half the globe. At sea were sand-banks, shoals, and whales; on shore, nothing was beheld but desolate plains and shaggy monsters of men, half-naked cannibals with claws, who lived in caves and holes in the sand, and in huts made of olive-branches or the ribs of fishes.*

The descriptions given of the Arabian Gulf, and of the manners of the inhabitants, are equally appalling. The coasts are represented as peopled with a race of savages, who fed on plants and leaves, dwelt in huts built on trees, and lived on fish, which they roasted on the rocks by the heat of the sun. They were expert in climbing, and could leap from branch to branch with great celerity. They excelled as marksmen and hunters, and caught elephants by cutting the trees nearly through against which they leaned to sleep. Their funerals they celebrated with mirth and dancing. When weary of life they strangled themselves, as well as their aged parents and infirm relations, by tying a bullock's tail round their necks.

All wonders naturally increase in proportion as they are distant and unknown; and it cannot be doubted that ignorance and imagination did much to enhance the difficulties and magnify the dangers of the Ara-

* "Omnino hæc Arabiæ continentis præternavigatio plena est periculi; regio impetuosa, infesta cautibus, atque scopulis inaccessa, horroris ubique plena."—(Arrian, Perip. Mar. Eryth. p. 12.) Pliny (lib. vii. cap. 24) speaks of the natives inhabiting these coasts as "hairy all over except the head, and clothed with the skins of fishes." Diodorus (lib. ii. iii.) describes the elephant-eaters, ostrich-eaters, fish-eaters, dog-eaters, locust-eaters, wood eaters, &c., dwelling near the Red Sea; as also men with cloven tongues, that spoke two languages at once.

bian seas. Avarice and speculations in commerce tended to aggravate these ideal horrors. The Romans and Egyptians had engrossed the trade of the East: willing to retain the profits of this intercourse themselves, and anxious to exclude foreigners from their ports, they cunningly spread exaggerated accounts of the perils to be encountered in visiting those mysterious regions from which they drew their wealth; as if nature herself, by the impenetrable deserts and oceans with which she had surrounded them, had set bounds to the cupidity of other mortals. The terrors of antiquity have been perpetuated in the modern nomenclature of the country. Hadramaut, the Land of Incense, means the Region of Death. The strait so formidable to the early navigators, and often indeed so fatal to their inexperience, the Arabs call Bab el Mandeb, or the Gate of Tears; while the opposite coast, black and rugged, they styled the Cape of Burials, on whose rocky steep their fancy heard the shrill spirit of the storm, as he sat in clouded wrath and enjoyed the death of the mariner. The light of science has dispelled these superstitious fancies. The navigation of that gulf is still intricate, its shoals numerous, and fatal mistakes are occasionally made; but nautical skill has rendered these disasters less frequent.

From the mouth of the strait to Gebel Tar, the soundings are from twelve to fifty fathoms, and there is a good landmark in the great mosque at Mocha. "The entrance to Bab el Mandeb," says Mrs. Lushington, "affords a sight equally unique and grand. A rush of the sea appears to have divided a bed of hard black rock, and thus to have forced a channel for itself of two or three miles in breadth. This rock rises on each side, dark, barren, and cheerless; that on the left is Perim; in some places a few blades of grass endeavour to force themselves through the crevices; but even fresh water must be brought from the Abyssinian shore,—the scarcity of this most necessary article being thus added to many other privations.* On this desolate spot Colonel Murray and a detachment of British troops were stationed during the French invasion of Egypt at the beginning of the present century. Lord Valentia has mentioned several errors discovered by Captain Court, who completed his survey in 1806. He states, that the actual distance between the island of Perim and the nearest part of Africa is only ten miles and a half, instead of sixteen, as laid down in the chart of Sir Home Popham; that the distance between the two shores, in latitude 13°, is only thirty-five miles, instead of fifty-two; and that there exists a shoal in that latitude which narrows the channel to fifteen miles, and is entirely omitted by Sir Home.†

To the ignorance and rude apparatus of the Arabs the Red Sea is still a dangerous passage. Not daring to venture into the open waters, the native pilots coast round the shores, at the hazard of being dashed in pieces upon jutting rocks, or stranded on coral reefs. Their ordinary vessels are dows and khanjas. The latter are large boats, but without any deck, save a little on the bows and that of the fore awning, under which is the cabin, open to the front, but without ports or windows; light and air being admitted through a neat open wicket at the side. This is all the improvement that has been made since the days of Arrian, who speaks of the small boats made of skins and plaited stuffs, or of single logs of wood called monoxyla, which the ancient inhabitants used for fishing pearls, employing osier baskets instead of nets.‡ The dows are of a singular construction;

* Journey from India, chap. ii. † Travels, vol. ii. p. 403.

‡ The rude structure of their boats in former times was more the effect of superstition than of ignorance. "It was an ancient prejudice of those frequenting the Erythræan Sea," says Procopius, "that rocks attracted iron, though the Roman ships found no such thing." Hence the Arabs carefully excluded that

their height, according to Ali Bey, being equal to a third of their length. The ropes are made of the bark of palm-trees, and the sails of extremely coarse cotton. To guard against the shoals, they have a false keel, which lessens the shock, and saves the ship if the weather is not rough. It is usual to cast anchor at night, except when crossing the gulf at its full breadth.

The vessel in which Niebuhr embarked at Suez for Jidda was large enough to have carried at least forty guns; and, besides her own freight, towed after her three large shallops and one small, the former being filled with passengers, horses, sheep, and women, belonging to the crews. On the appearance of a storm the sailors leaped into the boats and betook themselves to the shore. The pilot was constantly begging brandy of those on board, on pretence that he could not see the hills or the outline of the coast unless his sight was cleared by drinking a little strong liquor. On nearing the destined port their joy was excessive; cannons and muskets were fired, the ship and the boats were illuminated with lamps and lanterns; all was exultation and gratitude for the perils they had escaped

The superiority of European science has in a great measure set the impediments of rocks and winds at defiance. The British flag waves in every port of the Red Sea, from Suez to Aden. It has even been proposed to open a communication through that channel with India by means of steam, as much more expeditious than the ordinary passage by the Cape of Good Hope. Sir John Malcolm, in a paper on this subject which appeared in the Edinburgh Cabinet Library, states that for nine months in the year packets from India may be delivered at Alex-

dangerous metal from their naval architecture. The Romans forbade the sale of iron to the Indians and Ethiopians under pain of death.—*De Bell. Persic.*

andria in twenty-four days, and at Suez, with proper supplies of fuel, in three weeks. The voyage from Bombay to Aden (1640 miles) was performed in December 1830, in ten days and nineteen hours.*

TABLE OF GEOGRAPHICAL POSITIONS.

Names of Places	Lat. North. ° ′ ″	Long. East. ° ′ ″	Authorities
Mecca - - -	21 28 9	40 14 45	Ali Bey
Medina - - -	25 13 —	40 12 15	Ali Bey
Jidda - - -	21 32 42	39 5 45	Ali Bey
Yembo - - -	24 7 6	37 32 15	Ali Bey
Gonfode - -	19 7 —	41 42 —	Niebuhr
Loheia - - -	15 42 —	42 8 45	Niebuhr
Beit el Fakih -	14 31 17	— — —	Niebuhr
Taas - - - -	13 34 7	— — —	Niebuhr
Mocha - -	13 16 —	43 10 15	Con. des Tems
Sanaa - - -	15 21 —	44 — —	Neibuhr
Aden - - -	12 32 —	— — —	Niebuhr
Muscat - - -	23 38 —	59 15 —	Buckingham
Katif - - - -	26 20 —	49 50 —	Jomard
Yemama - -	24 — —	— — —	Jomard
Bahrein - - -	26 18 —	50 35 —	Sadlier
Deraiah - - -	25 15 —	46 30 —	Jomard
Bussora - - -	30 29 30	47 34 15	Buckingham
Petra - - -	30 20 —	— — —	Ptolemy
Ras Mohammed	27 50 —	— — —	Con. des Tems
Tor - - - -	28 12 19	33 33 10	Niebuhr

* British India, vol. iii.(a) It appears from the published correspondence which took place in October, 1832, between Mr. Barrow of the Admiralty and Messrs. Larpent and Begbie, the chairman and secretary of the East India Trade Committee, that the idea of this steam communication has been abandoned on account of the expense. But it is probable the project will be revived so soon as parliament comes to a final arrangement on the question respecting the East India Company's privileges. On this subject, to which we shall afterward advert, see "Remarks on the Advantages and Practicability of Steam Navigation from England to India," by Capt. C. F. Head; "Reports on the Navigation of the Euphrates," by Capt. Chesney; and "An Account of Steam-vessels, &c. in British India," by G. A. Prinseps.

(a) Vol. XLIX. Family Library.

CHAPTER III.

PRIMITIVE INHABITANTS OF ARABIA.

Obscurity of Arabian Antiquities—Want of written Records—Aboriginal Tribes—The old extinct Arabs—The pure Arabs—The mixed or naturalized Arabs—Their Attention to their Genealogies—Birth and Expulsion of Ishmael—Building of the Kaaba or Temple at Mecca—Death of Ishmael—Genealogy of Mohammed—The Koreish—Reflections on the National Descent of the Arabs.

The Arabian antiquities, like those of many other ancient countries, are extremely dark and uncertain. No nation, perhaps, whose history ascends without interruption to so remote an origin, or whose name has been so celebrated, has its political infancy enveloped in so thick a mist of doubt and oblivion. Shut up for so many ages within their rocky peninsula, they appear to have occupied themselves entirely with their own feuds and factions, which left them neither taste nor leisure for other avocations. Their chief study was a knowledge of their genealogies; but these could only preserve isolated facts: and, in so far as appears, they possessed no general annals,—no historical records, either common to the whole nation or to particular tribes. Songs and tradition perpetuated from one generation to another the superstitions and idolatries of their forefathers, the wars and exploits of their chiefs, and the invasions of their enemies.

In the absence of a national literature, it will not be surprising that we should find the narrative of those distant times so much corrupted by a mixture of absurd and improbable circumstances. Except a few monumental inscriptions and remains of poetry, a mass of traditions, disfigured by fiction and fable, is all that has escaped the oblivious wreck of these dark ages. We are apt to imagine that the zealous Moslems must, in the relentless spirit of their new creed, have swept away every record of the past, as infected with the errors of idolatry; and that the unsparing fanaticism, which proved so calamitous to arts and letters in other countries, had already committed a barbarous parricide on the ancient monuments of its own nation. This supposition, however, is not supported by any fact that has yet come to light. Some writers indeed, have asserted that, prior to Mohammed, historical annals and writings on different subjects existed. But as no such documents are to be found, or appear to have been consulted by the earliest Arabian historians, these assertions deserve little attention. On the contrary, the most ancient and learned among them agree in the confession that their old chronicles are traditional and imperfect; and that they could procure but indistinct notions of the times anterior to the Mohammedan era. All the authors extant or known in Europe who have treated of this period, such as Abulfeda, Hamza of Ispahan, Nuvairi, Masoudi, Al Tabiri, and Abulfarage, flourished after that epoch; and, except what we glean from the pages of sacred or Greek and Roman writers, it is from them we must derive our knowledge of the legendary ages that preceded the Saracen conquests.*

On one point there is a universal correspondence

* Collections from the works of these Arabic authors were translated and published (1786) by Albert Schultens, in the Historia Joctanidarum. The same editor has given specimens of their ancient language and poetry in his Monumenta Vetust. Arab. Ismael Abulfeda, prince of Hamah in Syria, a geographer and historian, died in 1345. Masoudi, author of the "Golden Meadows," an historical work, died in 957. Nuvairi, surnamed Al Kendi, author of a Universal History, died about 1340. Al Tabiri (a native of Tabreez), the Livy of the Arabians,

in their records,—that of their national descent. History and tradition agree in deducing their origin from Kahtan or Joktan, the son of Heber, and of the posterity of Noah by Shem. Among themselves this account has always passed as authentic. Elmacin calls Joktan the father of the Arabs; and Abulfeda adds, that his descendants inhabited Yemen, or the Happy Arabia. The parts of that country bordering on Palestine and Egypt were originally peopled by Cush, the son of Ham, whose descendants formed several petty monarchies and independent governments. Hence the name has been applied both by sacred and profane writers to Arabia as well as Ethiopia. Strabo, Diodorus, and Ptolemy peak of the Chusi and the island of Chutis as being in the former. The wife of Moses is called an Ethiopian, or native of Cush; but we know that she was an Arabian, and fed her father's flocks in the deserts of Horeb. In the prophecies of Habakkuk (iii. 7), Cushan and Midian are conjoined as the same territory. Sheba, Dedan, Teman, and other districts attest beyond dispute the names of the ancient settlers in these provinces. Various tribes have been already mentioned as occupying the borders of the desert from the Red Sea to the Chaldean Mountains, who were displaced by the posterity of Edom, to whom that region was a sort of promised land. But the Arabs pass them in total silence, as not sprung from either of the two acknowledged patriarchs of their nation. We shall therefore adhere to the following classification, which has been uniformly adopted by their own authors —The old extinct Arabs; the genuine or pure Arabs; and the mixed or naturalized Arabs.

finished his General History in 914. By the advice of his friends, he reduced it from 30,000 sheets to a more reasonable size. Price, in his Essay towards a History of Arabia, has given translations from it. Elmacin, whose Historia Saracenica was published in 1625 by Erpenius, is said to be abridged from Tabiri.

I. Of the old lost Arabs tradition has preserved the names of several tribes, as well as some memorable particulars regarding their extinction. This may well be called the fabulous period of Arabian history; but, as it has the sanction of the Koran, it would be sacrilege in a Mussulman to doubt its authenticity. According to this account, the most famous of the extinct tribes were those of Ad, Thamud, Jadis, and Tasm, all descended in the third or fourth generation from Shem. Ad, the father of his tribe, settled, according to tradition, in the great desert Al Akhaf soon after the confusion of tongues. Sheddad, his son, succeeded him in the government, and greatly extended his dominions. He performed many fabulous exploits: among others, he erected a magnificent city in the desert of Aden, begun by his father, and adorned it with a sumptuous palace and delightful gardens in imitation of the celestial paradise, and to inspire his subjects with a superstitious veneration for him as a god. This superb structure, we are told, was built with bricks of gold and silver alternately disposed. The roof was of gold, inlaid with precious stones and pearls. The trees and shrubs were of the same precious materials. The fruits and flowers were rubies; and on the branches were perched birds of similar metals the hollow parts of which were loaded with every species of the richest perfumes, so that every breeze that blew came charged with fragrance from the bills of these golden images. To this paradise he gave the name of Arem or Irem. On the completion of all this grandeur Sheddad set out with a splendid retinue to admire its beauties. But Heaven would not suffer his pride and impiety to go unpunished; for, when within a day's journey of the place, they were all destroyed by a terrible noise from the clouds. As a monument of Divine justice, the city, we are assured, still stands in the desert, though invisible.* The whole fable seems a confused tradition of Belus and the ancient Babylon; or rather, as the name would import, of Ben-hadad, mentioned in Scripture as one of the most famous of the Syrian kings, and who, we are told, was worshipped by his subjects.

Of the Adites and their succeeding princes nothing certain is known, except that they were dispersed or destroyed in course of a few centuries by the sovereigns of Yemen.

The tribe of Thamud first settled in Happy Arabia; and on their expulsion they repaired to Hajir (Petræa) on the confines of Syria. Like the Adites, they are reported to have been of a most gigantic stature, the tallest being a hundred cubits high, and the least sixty. And such was their muscular power, that, with a stamp of the foot in the driest soil, they could plant themselves knee-deep in the earth. They dwelt, the Koran informs us, "in the caves of the rocks, and cut the mountains into houses, which remain unto this day." In this tribe it is

* Southey, in his Thalaba, has viewed this, and many other of the fables and superstitions of the Arabs, with the eye of a poet, a philosopher, and an antiquary. According to Tabiri, this legendary palace was discovered in the time of Moawiyah, the first caliph of Damascus, by a person in search of a stray camel. To sum up the marvellous, a fanciful tradition adds, that the Angel of Death, on being asked whether, in the discharge of his inexorable duties, an instance had ever occurred in which he had not felt some compassion towards his wretched victims, admitted that only twice had his sympathies been awakened,—once towards a shipwrecked infant, exposed on a solitary plank, to struggle for existence with the winds and waves; and, a second time, in cutting off the unhappy Sheddad at the moment when almost within view of the glorious fabric which he had erected at so much expense. No sooner had the angel spoken than a voice from heaven was heard to declare that the helpless innocent on the plank was no other than Sheddad himself; and that his punishment was a just retribution for his ingratitude to a merciful and kind Providence, which had not only saved his life, but raised him to unrivalled wealth and splendour.—*Price's Essay*, p. 40.

easy to discover the Thamudeni of Diodorus, Pliny, and Ptolemy. The notion of giants is not uncommon, though it may in this instance have arisen from a mistranslation of the Koran. It is curious that the sons of 'Anak destroyed by Joshua (xi. 21) dwelt near the same place. The Jewish rabbis make Japhet and his son giants. The latter, they add, inherited an iron machine from his grandfather Noah, every stroke of which, when rightly aimed, slew a thousand men; and when not aimed at all it slew five hundred. The circumstance of dwelling in caves gave rise to the name of Troglodytes; and this was common to other tribes besides the Thamudites. Bruce observed them in Abyssinia; and Horneman found them in Fezzan, where they had existed since the time of Pliny.

The tribes of Tasm and Jadis settled between Mecca and Medina, and occupied the whole level country of Yemen, living promiscuously under the same government. Their history is buried in darkness; and when the Arabs wish to denote any thing of dubious authority, they call it a fable of Tasm.

The extinction of these tribes, if we may believe the Koran, was very miraculous, and a signal example of Divine vengeance. The posterity of Ad and Thamud had abandoned the worship of the true God, and lapsed into incorrigible idolatry. They had been chastised with a three years drought, but their hearts remained hardened. To the former was sent the prophet Hud (or Heber), to reclaim them, and preach the unity of the Godhead. "O people!" exclaimed the prophet, "understand and be converted, and supplicate remission for your sins! Then shall the heavens drop with rain, and your sustenance shall be renewed." Few believed; and the overthrow of the idolaters was effected by a hot and suffocating wind, that blew seven nights and eight days without intermission, accompanied with a terrible earthquake, by which their idols were broken to pieces, and their houses thrown to the ground.

Lokman, who, according to some, was a famous king of the Adites, and who lived to the age of seven eagles, escaped, with about sixty others, the common calamity. Those few that survived gave rise to a tribe called the Latter Ad; but on account of their crimes they were transformed, as the Koran states, into apes or monkeys. Hud returned to Hadramaut, and was buried near Hasec, where a small town (Kabr Hud) still bears his name. Among the Arabs, Ad expresses the same remote age that Saturn or Ogyges did among the Greeks; any thing of extreme antiquity is said to be "as old as King Ad."

The idolatrous tribe of Thamud had the prophet Saleh sent to them, whom D'Herbelot makes the son of Arphaxad, while Bochart and Sale suppose him to be Peleg, the brother of Joktan. His preaching had little effect. The fate of the Adites, instead of being a warning, only set them to dig caverns in the rocks, where they hoped to escape the vengeance of winds and tempest. Others demanded a sign from the prophet in token of his mission. As a condition of their belief they challenged him to a trial of power, similar to what took place between Elijah and the priests of Baal, and promised to follow the Deity that should gain the triumph. From a certain rock a camel big with young was to come forth in their presence. The idolaters were foiled; for on Saleh's pointing to the spot, a she-camel was produced with a young one ready weaned. This miracle wrought conviction in a few; but the rest, far from believing, hamstrung the mother, killed her miraculous progeny, and divided the flesh among them. This act of impiety sealed their doom. "Whereupon," adds the Koran (chap. vii.), "a terrible noise from heaven assailed them, and in the morning they were found prostrate on their faces and dead!" The caves inhabited by those infidel tribes rather militate against the idea of their gigantic stature, as the height does not exceed the ordinary standard. Their name and the places they inhabited are held accursed by all true Mussulmans.

The tribes of Jadis and Tasm owe their extinction to a different cause. A certain despot, Abulfeda relates, a Tasmite, but sovereign of both tribes, had rendered himself detested by a voluptuous law, claiming for himself a priority of right over all the brides of the Jadisites. This insult was not to be tolerated. A conspiracy was formed. The king and his chiefs were invited to an entertainment. The avengers had privately hidden their swords in the sand, and in the moment of mirth and festivity they fell upon the tyrant and his retinue, and finally extirpated the greater part of his subjects.

Besides those lost tribes others are enumerated, viz. Amalek, Abil, Waber, Jorham, Emim, and Jasim. All we know of them is, that they were either cut off in domestic feuds, or incorporated with other families.*

Such are the traditions regarding the extinct tribes of the ancient Arabs. History perhaps stoops from her dignity in noticing legends so fabulous and confused. The only importance they can claim is derived from being incorporated with the literature and religion of the country. Not only is much of the ancient poetry of the Arabs, their maxims, allusions, and proverbs, founded on them; but, what to us must appear still more absurd, the moral injunctions of the Koran, and the sacred title of the Prophet, are enforced by solemn references to the visionary punishments and idolatries of these defunct heretics. In the eye of a Mussulman those legends carry all the reverence of pious and indubitable

* Pococke, Specim Hist. Arab. p. 35. Sale's Koran, Prelim Diss. sect. i.

truths. They form the annals of his country,—the only remaining traces he has of its origin and history; and, however extravagant they appear, they unquestionably merit some attention. They mix with the national habits, and often influence the national character. The present generation have their faith strengthened and their duty taught by means of the fables of antiquity. Their bravery and their fanaticism are alike stimulated by popular traditions, which from their infancy they have learned to venerate.

II. The pure Arabs are those descended from Kahtan, whom the present Arabs regard as their principal founder. Like the Hebrews, a member of this genuine stock is styled Al Arab al Araba, an Arab of the Arabs. According to their genealogy of this patriarch, his descendants formed two distinct branches. Yarab, one of his sons, founded the kingdom of Yemen, and Jorham that of Hejaz. These two are the only sons spoken of by the Arabs. Their names do not occur in Scripture: but it is not improbable they were the Jerah and Hadoram mentioned by Moses as among the thirteen planters of Arabia.

In the division of their nation into tribes the Arabs resemble the Jews; and though, after the lapse of many thousand years, and over so vast an extent of territory, it is not to be presumed that each tribe could preserve an unbroken line of descent, yet their care and accuracy in this respect were remarkable. From the earliest era they have retained the distinction of separate and independent families. This partition was adverse to the consolidation of power or political influence; but it furnishes our chief guide into the dark abyss of their antiquities.

The posterity of Yarab spread and multiplied into innumerable clans. New accessions rendered new subdivisions necessary. In the genealogical tables of Sale and Gagnier are enumerated nearly three score tribes of genuine Arabs, many of whom became celebrated long before the time of Mohammed; and some of them retain their names even at the present day. Many Jews had settled in Arabia after the age of Joshua and Moses, where they formed powerful and independent tribes, and continued till the sword of the Prophet, their implacable enemy, either destroyed them or compelled them to abandon the country.

III. The third class are the Mostarabi, the mixed or naturalized Arabs, descended from Ishmael and the daughter of Modab, king of Hejaz, whom he took to wife, and who was of the ninth generation from Jorham, the founder of that kingdom. Of the Jorhamites, till the time of Ishmael, little is recorded, except the names of their princes or chiefs, and that they had possession of the territory of Hejaz. But, as Mohammed traces his descent to this alliance, the Arabs have been more than usually careful to preserve and adorn his genealogy. The want of a pure ancestry is, in their estimation, more than compensated by the dignity of so sacred a connexion; for they boast as much as the Jews of being reckoned the children of Abraham. This circumstance will account for the preference with which they uniformly regarded this branch of their pedigree, and for the many romantic legends they have grafted upon it. It is not improbable that the old giants and idolaters suffered an imaginary extinction to make way for a more favoured race; and that Divine chastisements always overtook those who dared to invade their consecrated territories.

The Scripture account of the expulsion and destiny of this venerated progenitor of the Arabs is brief, but simple and affecting. Ishmael was the son of Abraham, by Hagar, an Egyptian slave. When fourteen years of age he was supplanted in the hopes and affections of his father by the birth of Isaac, through whom the promises were to descend. This event made it necessary to remove the unhappy female and her child, who were accordingly sent forth to seek their fortune in some of the surrounding unoccupied districts. A small supply of provisions, and a bottle of water on her shoulder, was all she carried from the tent of her master. Directing her steps towards her native country, she wandered with the lad in the wilderness of Beer-sheba, which was destitute of springs. Here her scanty stock failed, and it seemed impossible to avoid famishing by hunger or thirst. She resigned herself to her melancholy prospects; but the feelings of the mother were more acute than the agonies of want and despair. Unable to witness her sòn perish before her face, she laid him under one of the shrubs, took an affecting leave of him, and retired to a distance. "And she went, and sat her down over-against him a good way off, as it were a bowshot; for she said, Let me not see the death of the child. And she sat over-against him, and lift up her voice, and wept." —(Genesis xxi. 16.) At this moment an angel directed her to a well of water close at hand,—a discovery to which they owed the preservation of their lives. A promise formerly given was renewed, that Ishmael was to become a great nation,—that he was to be a wild man,—his hand against every man, and every man's hand against him. The travellers continued their journey to the wilderness of Paran, and there took up their residence. In due time the lad grew to manhood, and greatly distinguished himself as an archer; and his mother took him a wife out of her own land. Here the sacred narrative breaks off abruptly,—the main object of Moses being to follow the history of Abraham's descendants through the line of Isaac. The Arabs, in their version of Ishmael's history, have mixed a great deal of romance with the narrative of Scripture. They assert that Hejaz was the district where he settled; and that Mecca, then an arid wilderness, was the identical spot where his life was providentially saved, and where Hagar died and was buried. The well pointed out by the angel they believe to be the famous Zemzem, of which all pious Mussulmans drink to this day. They make no allusion to his alliance with the Egyptian woman, by whom he had twelve sons (Gen. xxv. 12–18), the chiefs of as many nations, and the possessors of separate towns; but, as polygamy was common in his age and country, it is not improbable he may have had more wives than one.

It was, say they, to commemorate the miraculous preservation of Ishmael that God commanded Abraham to build a temple, and his son to furnish the necessary materials. By their joint labours the Kaaba or sacred house was erected, and solemnly consecrated by the Father of the Faithful, who prayed fervently that they and their whole race might become good Mussulmans. Its shape and substance were an exact type of Adam's oratory, which was constructed in heaven, and preserved from the deluge, to be a model to the venerable architects of the Kaaba. The black stone incased in the wall, and still pressed with devotion by the lips of every pilgrim, was that on which Abraham stood. It is alleged to have descended from heaven, and served him for a scaffold; rising and falling of its own accord, as suited his convenience. It was at first whiter than milk, but grew black long ago by the crimes or the kisses of so many generations of sinful worshippers.

The temple and the well became objects of general attraction. The Arabs conceived it a duty to adore Providence on the spot which bore such visible tokens of the Divine goodness. From the celebrity of the place, a vast concourse of pilgrims flocked to it from all quarters. Such was the commencement of the city and the superstitious fame of Mecca, the very name of which implies a place of great resort. Whatever credit may be due to these traditions, the antiquity of the Kaaba is unquestionable; for its origin ascends far beyond the beginning of the Christian era. A passage in Diodorus has an obvious reference to it, who speaks of a famous temple among the people he calls Bizomenians, revered as most sacred by all the Arabians.*

Ishmael was constituted the prince and first high-priest of Mecca; and, during half a century, he preached to the incredulous Arabs. At his death, which happened forty-eight years after that of Abraham, and in the 137th of his age, he was buried in the tomb of his mother Hagar. Between the erection of this edifice and the birth of their prophet, the Arabs reckon about 2740 years. Ishmael was succeeded in the regal and sacerdotal office by his eldest son Nebat; although the pedigree of Mohammed is traced from Kedar, a younger brother. But his family did not long enjoy this double authority; for, in progress of time, the Jorhamites seized the government and the guardianship of the temple, which they maintained about 300 years. These last, again, having corrupted the true worship, were assailed, as a punishment of their crimes, first by the scimitars of the Ishmaelites, who drove them from Mecca, and then by divers maladies, by which the whole race finally perished. Before quitting Mecca, however, they committed every kind of sacrilege and indignity. They filled up the Zemzem well, after having thrown into it the treasures and sacred utensils of the temple; the black stone; the swords and cuirasses of Kolaah; the two golden gazelles, presented by one of the kings of Arabia: the sacred image of the ram substituted for Isaac; and all the precious moveables, forming at once the object and the workmanship of a superstitious devotion. For several centuries the posterity of Ishmael kept possession of the supreme dignity. The fol-

* Ιερον αγιωτατον ιδρυται τιμωμενον ὑπο παντων Αραβων περιττοτερον.—Lib. iii. c. 3. Maximus of Tyre, in the second century, attributes to the Arabs the worship of a quadrangular stone (λιθος τετραγωνος).

lowing is the list of princes who swayed the sceptre of Hejaz, and who derive their chief fame from being the lineal ancestors of Mohammed:—

B. C.	122, Adnan.	A. D.	241, Galeb.
	89, Maad.		274, Lowa.
	56, Nazar.		307, Caab.
	23, Madar		340, Morra
A. D.	10, Alyas.		373, Kelab.
	43, Modreca.		406, Kosa.
	76, Khozaima.		439, Abdolmenaf.
	109, Kenana		472, Hashem.
	142, Nader.		505, Abdolmotalieb
	175, Malec.		538, Abdallah.
	208, Fehr.		

The period between Ishmael and Adnan is doubtful, some reckoning forty, others only seven generations. The authority of Abulfeda, who makes it ten, is that generally followed by the Arabs; being founded on a tradition of one of Mohammed's wives. It is not easy, however, to reconcile this discrepancy. Making every allowance for the patriarchal longevity of human life, even forty generations are insufficient to extend over a space of nearly 2500 years. But from Adnan to Mohammed the genealogy is considered certain; comprehending twenty-one generations, and nearly 160 different tribes, all branching off from the same parent stem.

The history of these petty sovereigns presents nothing memorable. Nazar, we are told, was a faithful adherent to the religion of Abraham. In his last will he made a singular distribution of his property. To Madar he bequeathed his red tent, and all his other effects of the same colour; to Rabia he assigned such of his moveables as were black; to Ayad the gray; and to Anmar the brown. The interpretation of this testament led to a display of Arabian sagacity, which is perhaps familiar to the reader in the tale of the Horse and the Dog, in Voltaire's Zadig; that being merely a transcript of an adventure that befell the sons of Nazar, who, from the minute description which they gave of a stray camel, were, on the mere strength of presumptive evidence, apprehended as guilty of theft. Fehr, surnamed Koreish, or the Courageous, was founder of the noble tribe of that name. That honourable title he obtained on account of his bravery in defending the temple; though the epithet would seem better merited by Kosa, who restored to the descendants of Ishmael the sovereignty of the city which had been usurped by the Khozaites, one of the emigrant clans from Yemen. Hashem surpassed all his predecessors in the grandeur and magnificence of his character. His generosity was unbounded; and hence the name Hashem, or the Divider of Bread. He killed vast numbers of his own camels to feed the indigent; his table was constantly furnished with all sorts of provisions, whether in times of plenty or distress, to which all were freely admitted. As the soil was too barren to produce a competent supply of corn and fruits, Hashem appointed two caravans to set out yearly,—the one in winter to Yemen and the south, the other to Syria. The commodities brought by these conveyances were distributed twice a year, and in such abundance that the poorest enjoyed all the luxuries of the rich. The vanity of the Arabs has exalted the glory and munificence of Hashem to the highest pitch. They even allege that the Roman emperor, hearing of his renown, sought an alliance with him, by offering him his daughter in marriage. This connexion was declined; for it would have been deemed sacrilege to mix the apostolical prerogative of the Koreish with the blood of strangers. The charities of the father descended richly on the son. Abdolmotalleb was liberal and hospitable; every year he entertained the poor on the flat roof of his house. So prodigal was his munificence, that he caused tables loaded with food to be transported to the summits of the mountains for the use of birds and wild beasts. He discovered, we may presume without the assistance of supernatural means, and restored, the treasures and other precious relics of the temple, which had lain buried in the Zemzem well since the expulsion of the Jorhamites. The swords he fabricated into an iron gate for the Kaaba; and this he gilded with the two gazelles, which he caused to be melted down, being the first gold with which that venerable edifice was adorned.

The national descent of the Arabs from Ishmael is a point which none will venture to dispute who receive the books of Moses as infallible authority. The historian of the Decline and Fall of the Roman Empire darkly insinuates his suspicions respecting the pedigree of this remarkable people. But the evidence of their derivation is too well established to be shaken by the efforts of the skeptic, who vainly thinks to invalidate the truths of Scripture by surrounding them with an air of fiction. That some uncertainty may have crept into their genealogies in the course of nearly thirty centuries will readily be allowed. But it is obvious that an isolated and unsubdued nation like the Arabs have the means of being more exact in the reckoning of their generations than in countries subject to changes and revolutions, where the pride of ancestry is necessarily obliterated and forgotten in course of a few successions. The lineage of Mohammed has been embellished with fables, and perplexed with anachronisms; but the veracity of Scripture, or the general interests of history, are in no respect impaired by circumstances so trifling.*

* Gibbon's Rom. Hist. chap. l. *note.* Brucker has arraigned the Scriptural genealogy of the Arabs. "Omnem quam Arabes *recentiores* jactant originem ab Abrahamo incertissimam esse."—*Hist. Crit. Philos.* tom. i. p. 214. The small number of generations mentioned in the long period of 2500 years is no solid objection. The Arabs did not always reckon from father to son.

One peculiarity by which the Arabs could distinguish the lineal ancestors of their apostle from all their collateral tribes was the extraordinary prophetic light that was said to illuminate their faces, —a symbol which had been inherited from father to son since the days of Adam. To him they allege it was communicated after his repentance, and imparted to none but the prophets; descending in the line of Seth, Noah, Shem, &c. After Abraham, it was separated into two parts, one remaining with the Jews, and the other with the Ishmaelites. All the progenitors of Mohammed bore this celestial imprint; faint or splendid, according to the faith and virtues of the individual. In some it was very largely developed. It did not, however, always follow the rule of primogeniture; and there is tolerable evidence that it enlightened some of very unsanctified habits. That the immediate predecessors of the Impostor were rich and powerful is unquestionable. By the establishment of caravans they had opened up new sources of wealth, and given trade a direction highly favourable to their own aggrandizement. We must, nevertheless, regard with caution many of the traditionary attributes ascribed to them, as they all bear the colouring of extreme partiality, and have tinged with romance the details of their private history.*

but from the heads of tribes; so that they preserved the purity of their descent without loading the memory with a catalogue of names. That the modern Bedouins are ignorant of their history is true; but they are not careless of their pedigree.—*Burckhardt's Notes on the Bedouins.* Forster has treated and established this point with great learning.—*Mahometanism Unveiled*, vol. ii. App. No. I.

* Pococke, Specim. Arab. Hist. p. 54. Gagnier, Vie de Mah tom. i. Introd. Rabadan, an Arragonian Moor, expatiates most extravagantly on the prophetic light, and traces the pedigree of Mohammed from the fall of Adam.—*Mahometanism Explained*, vol. i.

CHAPTER IV

ANCIENT KINGS OF ARABIA.

Subdivision of Arabian History—Want of Written Records—Defective Information of the Greeks and Romans—Confused Chronology of the Arabs—The Kings of Yemen, or Dynasty of the Hamyarites—The Flood of El Arem, and Destruction of Mareb—Exploits of Abucarb—Revolution under Dunowas—Persecution of the Christians—Invasion and Conquest of Yemen by the Abyssinians—Expedition of Abraha—War of the Elephant—Persians seize the Government of Yemen—The Kingdom of Hira or Irak—Kingdom of Gassan—The Nabathæan or Ishmaelite Arabs—Their Wars with the Jews and Romans—Expedition of Ælius Gallus—Perpetual Independence of the Arabs—Reflections on Gibbon's Skepticism—Recent Discovery of Petra—Description of its Magnificent Ruins.

The history of Arabia naturally divides itself into three periods, the Ancient, the Military, and the Modern. The first carries us down to the age of Mohammed, and is called by the Arabs the Times of Ignorance. The second includes the wars of the Saracens, and the empire of the caliphs. The third embraces the events from the fall of the caliphate to the present day. The native writers who treat of the first period all flourished, as has been observed, posterior to the era of the Prophet. It may seem remarkable that, among an intellectual and opulent people, no historians should have appeared to commemorate the events of their own times; but the causes are to be ascribed chiefly to their national character and habits. To the more civilized tribes the gains of commerce presented higher attractions than literary occupations; while the wandering hordes of the desert were content to devote the solitary hours of their monotonous life to the composition of songs, or the recitation of tales. Nor is it likely that a nation so proud of their independence would be careful to preserve their annals, when these could only record the invasions of their enemies, or an endless succession of domestic feuds, in which the weak constantly received the law from the strong. To have commemorated these inglorious transactions would only have been to perpetuate their own disgrace. It was, doubtless, from this impulse of national vanity, that no Arabian author has ever mentioned the presence of a Roman army in that country.

Little light is thrown on this obscure epoch from foreign sources, except at distant intervals. If we consult the Greek and Roman authors, the information they furnish is far from being exact. Strabo assures us, that Arabia the Happy was divided into four distinct governments, and that the succession of their kings was not fixed by primogeniture or even by royal descent. The right extended to a certain number of privileged families, and the first male child born after the commencement of a new reign was considered heir to the crown. Agatharcides, on the contrary, tells us that their kings were hereditary, and that so long as they remained shut up in their palaces they were greatly respected; but that the people assailed them with stones the moment they appeared in public, and after their death buried them in dunghills. Diodorus has recorded the same peculiarity. Arrian places one kingdom in the western part of Yemen and another to the eastward; the capitals of which he calls Sabbatha and Aphar, evidently the Saba and Dhafar of later writers.

One main difficulty resulting from the want of native or contemporary written records is, that of determining with any tolerable precision the chronological succession of events; for it does not appear that among the Arabs in the Times of Ignorance any particular era was generally adopted. In Yemen, where a regular sovereignty was so long maintained, it is somewhat remarkable that a more exact chronology should not have been observed. Several lists of these ancient kings have been preserved. We are told that they assumed the general name of Tobbaa, a title equivalent to Cæsar or Pharaoh among the Romans and Egyptians; but we know little about the nature of their power or their system of administration. This monarchy, according to Jannabi, extended to 3000 years, while Abulfeda restricts it to 2020. But how twenty-six or thirty kings could occupy even the shortest of these periods it is difficult to conjecture. The Mohammedan historians solve the perplexity by making some of them reign three or four hundred years, and live to nearly twice that age. "God only knows the truth!" is the constant exclamation of the pious Nuvairi, on finding it impossible to reconcile these computations with the ordinary limits of mortality. We rather agree with Pococke and M. de Brequigny,* that it was only those princes who swayed the undivided sceptre of Yemen, or were conspicuous as tyrants or conquerors, whose names have been preserved; and that the intervals, being filled up with usurpations, or not marked by any memorable events, have been passed over in studied silence. Hamza says expressly, that the twenty-six kings who flourished for so long a period were only those descended from the family of Hamyar.

Besides that of Yemen, there were two other principal dynasties in Arabia, of which we shall give some account in the following order:—

I. The kingdom of the Homerites or Hamyarites, so called from the fifth monarch of that name, who

* Mem. de l'Academ. des Inscript. tom. xxix. Spec. Hist. Arab. p. 55.

possessed the whole or the greater part of Yemen, the several petty princes who reigned in other districts being mostly, if not altogether, dependent on this sovereign, whom they called the Great King.

II. The kingdom of Hira, or the Arabian Irak, whose capital stood on the Euphrates.

III. The kingdom of Gassan on the borders of Syria. Its sovereigns were a kind of viceroys to the Roman emperors, as those of Hira were to the monarchs of Persia.

Kahtan, the founder of their race, is honoured by the Arabs as the first that wore the crown of Yemen.

Yarab, his son, they regard as the first that spoke their language. Saba built the capital called after himself; and hence the inhabitants got the name of Sabæans. Tables of these kings have been drawn up by various historians; but they differ so much in their calculations as to satisfy us that they are not to be trusted as infallible guides. Those given by Pococke have been generally followed, as being more complete, and at the same time more consonant with probability, than any to be found in a single Mohammedan author:—

I. TABLE.—Kings of Yemen,—Reigned 2020 years

1. Kahtan.	16. Dulkarnain.
2. Yarab.	17. Dulmenaar
3. Yashab.	18. Afreikus.
4. Saba (or Abd-Shems).	19. Duladsaar.
5 Hamyar.	20. Shaerhabil.
6 Wathel (or Wayel).	21. Hodhad.
7. Secsac.	22. Belkis.
8 Yaafar.	23. Nashirelnaim.
9. Duryash.	24. Shamar—Yaraash
10. Nooman.	25. Abimalec.
11. Asmah.	DESCENDANTS OF CAHLAN.
12. Sheddad.	26. Amran.
13. Lokman.	27. Amru—Mazikia.
14. Dusadad.	28. Akran.
15. Hareth,—Alrayish	29. Duhabshan

The history of these ancient kings is little else than a mere register of names. On the death of Hamyar, the family of his brother Cahlan disputed the throne, and divided the monarchy; one branch continuing to reign at Saba, and the other at Dhafar in Hadramaut. After a lapse of fifteen generations, these were united in the person of Hareth, surnamed Alrayish, or the Enricher, from the abundance of spoils he collected in his various expeditions. Having recovered the entire sovereignty of Yemen, he assumed the title of Tobbaa, or Successor. Dulkarnain, who has been erroneously identified with Alexander of Macedon, is a celebrated personage in oriental story. He pushed his conquests to the remotest regions of the earth, vanquished nations of colossal stature, and subdued towns whose walls and towers were of brass and copper, so brilliant that the inhabitants were obliged to wear masks to protect them from total blindness. This apocryphal prince is mentioned in the Koran (chap. xviii.), but it seems doubtful to what character in real history his achievements are to be ascribed. They certainly bear some resemblance to the romantic exploits of the all-subduing son of Philip. Dulmenaar, his successor, carried his arms westward into the unexplored regions of Nigritia, where he is said to have constructed a chain of lighthouses over the desert to guide his march; hence his name, which means Lord of the Watchtowers. His son extended his conquests as far as Tangier, and is said to have given his name to Africa. Duladsaar, or the Lord of Terror, is renowned as the conqueror of the Blemmyes or Pigmies, a nation of monsters without heads (Acephali), and having eyes and mouths in their breasts, whom Herodotus and Mela placed in Abyssinia and Southern Africa. His subjects threw off their allegiance, and raised Shaerhabil, a descendant of Wathel, to the throne, who, after several bloody battles, became undisputed master of the kingdom.

Belkis, according to the Arabs, was the famous Queen of Sheba or Saba, who visited and afterward married Solomon in the twenty-first year of her reign. Tabiri has introduced her story with such gorgeous embellishments, as to resemble a fairy tale rather than an episode in serious narrative. She is said to have been subdued by the Jewish monarch, who discovered her retreat among the mountains between Hejaz and Yemen by means of a lapwing, which he had despatched in search of water during his progress through Arabia. This princess is called Nicolaa by some writers. The Abyssinians claim the distinction for one of their queens; and have preserved the names of a dynasty alleged to have been descended from her union with Solomon.* Yasasin, surnamed Nashirelnaim, or the Opulent, from his immense wealth, has the reputation of being a magnificent and warlike prince. His ambition carried him into the unknown deserts of the West; but the whirlwinds of moving sands compelled him to return, after losing a great part of his army, which he had rashly ordered to advance. To commemorate this disaster, he caused a brazen statue to be erected on a pedestal of stone, with an inscription in the Hamyaric character, importing that here was the limit of his progress; and that none, but at the peril of destruction, could attempt to go beyond it. The military achievements of Shamar, called Yaraash, or the Tremulous, from a disease to which he was subject, resemble those of his predecessors. He is recorded to have made various expeditions to Persia. He subdued Khora-

* Russell's Nubia and Abyssinia, Family Library, No. LXI. From her designation of "Queen of the South" (Yemen), and the description of her presents to Solomon, "gold and spices, very great store," there is little doubt that Arabia was the native country of this famous princess.

san and other provinces; and, traversing Sogdiana, he laid siege to the capital, which he completely destroyed. From him Samarcand is alleged to have taken its name, according to an inscription said to have been engraven on one of the gates. This monarch with his whole army perished by a stratagem while attempting to penetrate the desert towards Chinese Tartary, of which he meditated the conquest. On the death of Abimalec the throne of the Hamyarites was usurped by the descendants of Cahlan; and accordingly the two brothers, Amran and Amru, are not recognised by some historians as kings of Yemen. They are omitted in the lists of Hamza, Nuvairi, and Masoudi. Amran was noted for his skill as a soothsayer; and Amru acquired the nickname of Mazikia, or the Tearer, because in his time the kingdom was divided; or, as others say, from the strange caprice he indulged of tearing his robes every evening, disdaining either to wear them again himself or allow others to do so.

Though the annals of the preceding dynasty are doubtless blended with romance, there seems good reason to believe that some of these Arabian monarchs were both enterprising and powerful. It would be useless, however, to form conjectures as to the reality or extent of their conquests; or attempt, at this distance of time, to reconcile the order of their succession with our systems of chronology. Kahtan, according to the Mosaic genealogy, was born 532 years after the flood; five or six generations, at the average rate of human life in those early ages, will bring us down to the death of Abraham; and this computation agrees with that of Nuvairi, who makes Hamyar coeval with Kedar, the son of Ishmael (B. C. 1430). Afreikus is said to have been contemporary with Joshua; but this supposition can hardly be reconciled with the statement that Belkis reigned in the days of Solomon (B. C. 901) As the dynasty of the Hamyarites changed with Abimalec, who is reckoned contemporary with Alexander the Great, this circumstance may with great probability account for the chronological blank that occurs between the time of that prince and the Christian era.*

The reign of Akran forms a memorable epoch in Arabian history, on account of the political changes alleged to have been occasioned by the flood of El Arem. The Mohammedan writers dwell at great length on this catastrophe, mixing its details with a variety of fabulous circumstances. The territory of Saba, though naturally fertile, had, according to Nuvairi and Masoudi, who have written elaborate treatises on this famous deluge, been rendered almost uninhabitable from the impetuosity of the mountain torrents, which destroyed their houses, harvests, and vineyards, and the whole produce of their industry. With a view to oppose some barrier to these ruinous floods, one of their kings, Saba, or Lokman, constructed a huge mole or bank, stretching across the valley, which was about a quarter of a mile in breadth, at the lower extremity of the adjacent mountains. It was built of solid masonry, the blocks of marble being cemented with bitumen and strengthened with iron bars. It rose to a great height above the city (Mareb), and was by the Sabæans deemed so strong, that many of them had their houses erected on its sides. The valley, to the distance of about five leagues, was thus converted into a vast lake 120 feet in depth, and receiving, according to Abulfeda, the tributary waters of seventy streams, some of which were conducted into

* The annals of Persia present a similar chasm. "From the death of Alexander till the death of Artaxerxes is nearly five centuries; and the whole of that remarkable era may be termed a blank in Eastern history. Yet, when we refer to Roman writers, we find this period abounds with events of which the vainest nation might be proud."—*Malcolm's Hist. of Persia*, vol. i. chap. iv. p. 68.

it by artificial channels. In this mound were thirty sluices at three different heights, each about a cubit in diameter, through which the waters issued, and were conducted with the aid of machinery through smaller canals to the fields, gardens, and houses of the inhabitants. Mareb thus became, as Pliny calls it, the mistress of cities, and a diadem in the brow of the universe.

This golden age of Arabian antiquity is a favourite theme with their poets and historians, who expatiate on the extensive fields and forests of Saba, its beautiful edifices, and numerous orchards A good horseman, says Masoudi, could scarcely ride over the length and breadth of this cultivated country in less than a month; and the traveller might wander from one extremity to the other without feeling the heat of the sun; for the thick foliage of the trees afforded a continual shade. Its luxuries were proverbial,—a pure air, a serene sky, wealth without its cares and inconveniences, all conspired to render Mareb the retreat of every blessing that can make life agreeable. The happy natives enjoyed among their groves and vineyards a peaceful and palmy security, clothed in embroidered garments of green silk, and rewarded with a double increase of their flocks and their fields. The kings were virtuous like their subjects. Their dominion, mild and equitable at home, was acknowledged and respected by the surrounding nations; for no enemy assailed them whom they had not defeated, and every region which they invaded had submitted to their arms.

The capital itself, we are gravely told by a Turkish geographer, was distinguished by twelve peculiarities, not less attractive than its abundant streams and delicious fruits. Neither serpents, flies, nor other troublesome insects were to be found in it: strangers infested with vermin, particularly the third plague of the Egyptians, no sooner entered it than they were relieved: none of its citizens were liable to sickness or disease ; the sick, the blind, the maimed, the paralytic, from other quarters, might all be restored by bathing in its waters : no change of dress was necessary, such was the mildness of the climate : their wives knew not the pangs of childbirth, and never lost the charms of youth and virginity.

All these imaginary felicities, however, depended on the strength and preservation of their mound; the effect of time (if built by Saba, it must have stood above 1700 years) and the weight of the water began insensibly to undermine its foundations. The king was apprized of the danger by Amru Mazikia, and a noted female soothsayer, Dharifa, an interpreter of dreams and visions, who announced, by many terrible signs and prodigies, the approaching devastation of Mareb. The incredulous prince disregarded every admonition; but Amru, having disposed of his property, resolved, with a number of followers, to seek safety in a timely flight. The bursting of the waters immediately overwhelmed the country, destroying fields, flocks, vineyards, and villages; and reduced that fertile province to a state of desolation. Such is the history of the famous deluge of El Arem.

As the Sabæans were a proud and idolatrous race, the Koran (chap. xxxiv.) describes this disaster as a judicial punishment from Heaven.—" Wherefore we sent against them the inundation of El Arem, and we changed their double gardens into gardens producing bitter fruit,—this we gave because they were ungrateful." The Arabian poets lamented its destruction in verse; and two elegies on the subject have been preserved among the ancient monuments of their literature. The tradition still exists among the inhabitants of Yemen, one of whom described to Niebuhr the ruins of the wall on the sides of the two mountains. The tributary waters had dwindled away to six or seven petty streamlets, some of which contained fish; but their course was speedily absorbed in the sands. The Danish traveller remarks, it would be a profitable and not a very difficult task for the government to rebuild the structure; but the local authorities are too poor to attempt the undertaking. Reservoirs similar to that of Mareb, on a smaller scale, are not uncommon in Yemen and other parts of the East. Those that supply Constantinople are constructed after the same manner. Abulfeda mentions one near Emesa (Homs) in Syria, which the natives attribute to Alexander the Great; and another at Tostar, in Persia, which raised the water of a neighbouring river to the level of that city. Tavernier and Chardin have described one very much resembling that of Saba, built at the extremity of a pleasant valley near Ispahan, for the purpose of collecting the rills and melted snows from the surrounding mountains.

Whatever truth there may be in the allegorical details of this catastrophe, there is no reason to doubt the event itself. That political causes, arising from civil war or foreign invasion, as much as the decayed state of the embankment, may have contributed to the revolution which then took place, is more than probable; but we can hardly suppose, as some have thought, that this deluge was a fiction of the Arabs to save their national reputation; since the occurrence of such a calamity is uniformly attested, both by their sacred and historical records.

No point in Arabian chronology has been disputed with more learning than the date of this inundation. Most of the native historians have fixed it about the time of Alexander the Great; but little credit is due to their loose calculations. They all agree, however, that it happened in the reign of Akran, or his son Duhabshan. Reiske places it 30 or 40 years, and De Sacy 140, after the Christian era. The former, notwithstanding the high autho ity against us, we are inclined to regard as the mo probable epoch. A considerable time must necessarily have elapsed after the overthrow of the capital, and the devastation of the country, before the scattered tribes could be again united, or the government consolidated under a single monarch. During this interval several petty princes appear to have reigned over these districts. The Roman historians mention Cholebus, whom they style king of Maphartis, and Charibael, whose residence was at Saphar, to whom some of the Cesars addressed embassies, and sent valuable presents, with a view to conciliate their friendship. In the course of little more than a century the throne of the Hamyarites was again firmly established in Yemen, by a descendant of Akran, who assumed the title of Tobbaa I. From this time the Arabian history is much more exact, as the reigns of the different princes are found to synchronize in most instances with those of the Persian monarchs. In the following table we adopt the chronology given in the learned Dissertation of the Baron De Sacy, who is, beyond dispute, of all living authors, the most profoundly conversant with the ancient literature of Arabia.*

II. Table.—Kings of Yemen,— Reigned A. D. 175–529.

A. D. 175,	Tobbaa I.	A. D. 321,	Morthed.
190,	Colaicarb.	345,	Wakia.
220,	Asaad-Abucarb.	370,	Abrahah.
238,	Hassan.	399,	Sahban.
250,	Amru-Dulawaad.	440,	Sabbah.
271,	Four anonymous kings.	455,	Hassan II.
272,	Alsaha.	478,	Dushanater.
273,	Abd-Kelal.	480,	Dunowas.
297,	Tobbaa-el-Asghar.	528,	Dujadan,
313,	Hareth.	529,	Dujazen.

After the destruction of Mareb, the power of the Tobbaas soon rose to more than its ancient splen-

* Mém. de l'Acad. des Inscrip. tom. xlviii. p. 544. Reiske de Arab. Epoch. Vetust. M. Gossellin adopts the epoch of Reiske. Recherches sur la Géographie des Anciens.

dour. Asaad-Abucarb had larger armies, and extended his conquests more widely than any of his predecessors. He invaded Tehama, the inhabitants of which were glad to purchase peace from him at the expense of twenty camels for every soldier they had slain. Carrying his arms eastward, he proceeded by the route of Mosul into Azerbijan, where he encountered and defeated the Tartars with great slaughter. Alarmed at his success, most of the neighbouring monarchs courted his friendship; and among these was the sovereign of Hindostan, who sent an embassy proposing terms of amity. The rare articles presented by the ambassador led to inquiries respecting the country which produced them; and for the first time the Arabian conqueror heard of the existence of China. Asaad at once determined on an expedition to that distant region; and quitted Yemen at the head of a force which oriental hyperbole has magnified into a thousand standards, each followed by a thousand men. Having by some means led his army through the territory of Balkh, he proceeded by Turkistan, skirting the borders of Thibet, where he left a division of 12,000 Arabs, as a body of reserve, in case of defeat. Finally, he succeeded in penetrating the boundaries of the Chinese monarchy; and, after plundering the cities in all directions, he returned with immense booty through Western Tartary into India, whence he conducted his army safely back to Yemen, having consumed seven years in this remote and perilous enterprise. The corps of reserve, however, was never withdrawn from Turkistan and Thibet, where vestiges of the race are still to be discovered.*

The whole of this expedition we might have been apt to treat as a fable, were it not that the early Mohammedans, on conquering Bokara, found an inscription expressly recording the presence of the

* Price (Essay, p. 98) is disposed to place this prince earlier

Hamyarite Tobbaa in that quarter. Harassed with his endless wars, which procured him the name of Abucarb, or the Father of Affliction, his subjects conspired his death, and transferred the crown to his son Hassan, who was himself slain by his brother Amru. This state of affairs led to an insurrection, in which the usurper with his four sons and their sister Alsaha were cut off; although the latter are not admitted by some among the number of his successors. Abd-Kelal, according to Hamza, embraced Christianity; but he was deterred by political motives from avowing it openly, or imposing it on his subjects. Tobbaa, the last that was honoured with that title, marched with an army of 100,000 men into Hejaz, threatening to exterminate the Jews, who had put his viceroy to death on account of his cruelties. A colony of these people had fled from Palestine and Syria, in the wars of Titus and Adrian, and settled near Yatreb (Medina). This city Tobbaa besieged; but a reconciliation having taken place, it was saved from destruction. Two Jewish doctors, it is alleged, had succeeded in impressing him with the danger of violating a place which was under the special protection of Heaven, and destined to become the future asylum of a great prophet. A similar veneration was the means of protecting Mecca and its temple from pillage. In reverence for its antiquity and holiness he presented the Kaaba with its first canopy, a cloth of rich tapestry, and a gate of gold; and during the six months of his residence there, he is said to have sacrificed every day a thousand camels. His intercourse with these Hebrew exiles led to a change in his religion. The doctrines they unfolded to him appeared so acceptable, that he instantly abandoned the absurdities of idol-worship, and became a zealous convert to the Mosaic ritual.

On his return to Yemen, he was accompanied by a number of Jews, whom he soon advanced to places of trust and authority. His own subjects, however opposed the introduction of a strange religion, and resolved no longer to acknowledge as their sovereign a prince who had deserted the faith of his ancestors. The dispute was at length adjusted by an appeal to a subterraneous fire, in a cavern near Sanaa, to which the people had been accustomed from time immemorial to submit all nice points of difference, wherein it was found impossible by ordinary means to discriminate right from wrong, or truth from falsehood. This infallible ordeal, as was to be expected, decided in favour of the Jewish rabbis, who entered the grotto with portions of their Scriptures suspended from their necks, and returned unhurt by the flames; while the idols of Yemen, and all those by whom they were carried, were instantly consumed to ashes. In consequence of this awful manifestation, the whole inhabitants, according to the Arabian legends, embraced the Law of Moses.

The history of this prince, the last of the Tobbaas, is very contradictory, many of his achievements being ascribed to Asaad-Abucarb, who is by some called the second or middle Tobbaa. His death, and the minority of his three sons, gave rise to scenes of turbulence and usurpation, that distracted the kingdom for many years.

The reigns of Dushanater and his successor were marked with infamy, and ended in revolution. The former, a tyrant noted for his vices and barbarities, was not of the royal lineage, but had seized the throne in the absence of Hassan II., then on an expedition to Syria. The usurper had succeeded in establishing his authority, by cutting off all whose hereditary claims stood in his way. His loathsome propensities made his government as odious as it was cruel. It was his practice to allure the sons of the nobility to his palace at Sanaa, and after subjecting them to the most brutal treatment, to hurl them from an upper window in the presence of his guards, who enjoyed with their master this revolting spectacle. Such of the royal progeny as had escaped the dagger were in this manner sacrificed to an infamy worse than death. One prince of the blood only remained,—Zerash, a youth of surpassing beauty, whose flowing locks obtained him the name of Dunowas. This last victim of his fallen race was seized and conducted to the fatal pavilion; but it was to revenge instead of sharing the ignominy of his unhappy kindred. Having secreted a small poniard under the sole of his foot, he contrived to stab the licentious tyrant to the heart. Severing the head from the body, he exhibited it at the window from which he was himself to have been precipitated. The court satellites gazed a moment in doubt and astonishment. Dunowas pointed to the bloody trophy as the best interpreter of what had taken place. The deed was hailed with applause, and the vacant diadem unanimously conferred on their deliverer.

This intrepid youth became one of the most formidable and powerful monarchs of Yemen, though his cruelty soon blighted the auspicious prospects with which his reign commenced. His bigotry to the Jewish faith, which led him to assume the name of Yussuf, rendered him an intolerant persecutor. The Christians especially felt his severity. The inhabitants of Nejeran, the Beni-Thaleb, who had been converted by a Syrian called Akeimoum, and had a bishop of their own, were doomed to indiscriminate extermination. Refusing to abjure their creed, they were thrust into a pit or trench filled with combustibles, to which burning fagots were applied, and in this manner 20,000 of them perished. The Lord of the Burning Pit is the terrible title which this inhuman act procured for Dunowas. The fidelity of the martyrs, or "brethren of the pit," is commended in the Koran, where an anathema is pronounced on their persecutor (chap. lxxxv.).

One of the few Christians that escaped applied for revenge to the Nayash or King of Abyssinia, who was a Christian, urging him to undertake the invasion of Yemen. An army of 70,000 men was accordingly despatched, under the command of his son Aryat, with injunctions to put to death every Jew, to pillage a third part of the country, and take captive a third part of the women and children. On landing at Aden, Aryat burnt his ships,—a signal to his troops that they must conquer or perish. Weakened by dissensions, and taken unawares, the Arabs were routed with great slaughter. Dunowas fled, and finding himself pursued, he spurred his horse to a rocky precipice, and threw himself into the sea: preferring a watery grave to the chains of the Ethiopian victor. The vengeance of the Christians thus proved fatal to the independence of Yemen. Two princes of the Hamyarite line, Dujadan and Dujazen, made an unsuccessful struggle to regain the sovereignty; but that ancient dynasty had lost the sceptre for ever, which was now transferred to the hand of an Abyssinian conqueror. Such is the account which the Arabian authors give of this famous invasion. The chronicles of Abyssinia, however, distinctly mention an earlier expedition across the Red Sea. The inscription at Axum, discovered by Mr. Salt, records that the reigning monarch there, Aizanas, had sent his two brothers into Arabia prior to the year 327, who subdued the Homerites, with several other tribes, and carried away a great number of captives, with their sheep, oxen, and beasts of burden; and these he established as a colony, at a place called Matara, in his new dominions. The final subjugation of Yemen is that referred to in our narrative, and is placed by the chronology of both countries in the time of Anastasius and his successor Justin. Cosmas, who visited Adulis in the reign of the latter emperor, states that Elesbaan or Caleb Negus (the navash of the

Arabs), was then on the point of undertaking an expedition into Arabia; and from this period must be dated the extinction of a race of princes, who had ruled above two thousand years the sequestered region of myrrh and frankincense.*

III. TABLE.—Abyssinian Kings of Yemen,—A. D. 529-601.

A. D. 529. Aryat or Arnat.	A. D. 589, Yacsum.
549, Abraha—Al Ashram.	601, Masruk.

As a condition of his victory, Aryat was confirmed in the government of Yemen; but the turbulent and artful policy of Abraha, an officer in the expedition, and formerly the slave of a Roman merchant at Adulis, shortened his reign. Supported by a part of the army, he revolted and offered battle. Instead of hazarding a civil war, it was agreed to decide the contest by single combat, in which the Abyssinian was treacherously stabbed by a slave; not, however, before he had wounded his antagonist in the face, which gave him the surname of Al Ashram, or the Slit-nosed. The nayash threatened to punish the rebel; and made a vow to drag him from his throne by the hair, to trample his dominion under foot, and die his spear in his blood. Abraha seems to have paid little regard to these menaces; and took an ingenious plan to accomplish their fulfilment without danger to himself. He filled two sacks with earth, cut off two locks of his hair, which, with a small vial of his blood, he enclosed in a rich casket perfumed with musk, and despatched to his master; expressing a hope that the royal displeasure would be satisfied with this easy mode of punishment, as he had thus given him an opportunity of executing his threat to the letter, without violating his conscience, or incurring the hazard and expense of an expedition. The nayash was

* Vincent's Perip. vol. i. App. ii.; Valentia's Trav. vol. iii. ch. vi.

pacified, and the usurper confirmed in his new dignity.

Being of the Christian profession, his efforts were directed to the conversion of his subjects. He built, among other structures, a splendid church in his capital; of such magnificence, we are assured, that it had no equal at that time in the whole world. "A huge pearl," says Nuvairi, who wrote a treatise on Arabian edifices, "was placed on the side of the altar, of such brilliancy that in the darkest night it served the purpose of a lamp." The object of Abraha was to make Sanaa the Jerusalem of Arabia, the holy city, where all pilgrims were in future to resort instead of Mecca. This kindled the indignant zeal of the idolaters, more especially of the hereditary guardians of the Kaaba, who saw in the popularity of the Christian temple the downfall of their own greatness. On the night of a solemn festival, two of the Koreish entered the church, and having profaned it by an act of gross indecency, fled to Mecca. Abraha vowed a terrible retaliation, declaring that not a stone of that obnoxious city should remain upon another. Forty thousand men were levied, of whom he took the command in person, riding on a white elephant of prodigious size and beauty. He routed in a single battle the inhabitants of Tehama, who had refused to transfer their religious allegiance to Sanaa. At Taif he ordered all the cattle of the surrounding districts to be seized, among which were two hundred camels belonging to Abdolmotalleb, prince and pontiff of Mecca.

The appearance of this formidable expedition before the sacred city spread general consternation, for the Meccans were neither able nor prepared for defence. Abdolmotalleb repaired to the camp of Abraha, where he was received with every mark of honourable distinction. "I come," said he, "to demand restitution of my cattle." "Why not," replied the invader, "rather implore my clemency in favour of your temple, the source of your grandeur, and the object of your religious veneration?" "The camels are my own," was his reply, "the Kaaba belongs to the gods, and they will defend it. Many kings have attempted its destruction, but their ruin or repulse has protected it from sacrilege." The camels were restored; but the temple was left to the protection of its own sanctity. The venerable pontiff retired with the citizens to the mountains and fortresses in the vicinity; having supplicated in a pathetic hymn, before their departure, that the calamity intended for the asylum of their faith might be visited on the heads of its enemies.

The deities of the place, if we are to believe the Arabs, were not importuned in vain; and by their interposition the Christian host met with a signal overthrow. Abraha advanced on his huge elephant Mahmoud; but neither violence nor entreaty could made it enter the consecrated walls. In any other direction, towards Syria or Yemen, it would move with the greatest alacrity; but not a single step towards the Kaaba. The other elephants, thirteen in number, evinced the same reluctance to commit sacrilege, and always knelt down, when turned to that quarter, in the same reverential attitude. A miracle at length relieved the city. An innumerable army of birds from the seacoast, like a dense cloud, suddenly appeared, hovering over the Abyssinians. Each carried a stone in its bill, and one in each claw, about the size of a lentil; these they let fall on the heads of the besiegers with such violence as to pierce through their helmets and armour, killing men and elephants on the spot. On every stone was inscribed the name of its particular victim. These birds, called Abil, are represented of a size between that of a swallow and a pigeon, and party-coloured, being black and white interspersed with green and yellow. The few invaders that escaped this supernatural catastrophe perished in the desert Abraha alone reached Sanaa, "quaking like a chicken," where he died soon after of a loathsome disease. The terrified citizens of Mecca returned from their hills. Abdolmotalleb and the Koreish were regarded with double veneration, and invested with the title of the Holy Family.

The War of the Elephant is a well-known epoch in Mohammedan history, as it happened in the year of the Prophet's birth. The Koran in a short chapter (cv.) relates this judicial defeat of "The Masters of the Elephant" by a miraculous flock of birds, "which cast down upon them stones of baked clay." It is difficult to comprehend how a legend so ridiculous, and happening at a period so well ascertained, could have gained the slightest degree of credit Dean Prideaux considers the story a fiction of Mohammed's own coining, to terrify the Christian Arabs into his religion, and augment the national reverence for the Kaaba. Father Maracci alleges the whole either to be a fable, or a feat of some evil spirit, such as overthrew Brennus and his army when marching to attack the temple of Apollo at Delphi Stripped of its preternatural absurdities, this memorable event will resolve itself simply into a religious expedition against Mecca; and the discomfiture of the Abyssinians, as Gibbon remarks, may be attributed either to the want of provisions, or to the valour of the Koreish, without the assistance of a celestial shower of stones.

Abraha was succeeded by his sons, Yacsum and Masruk. Their debaucheries and oppressions alienated the loyalty of the Arabs, and raised a competitor in the person of Seiph, a descendant of the last of the Hamyarite princes. He applied for aid to Khoosroo (Chosroes), king of Persia, whose wealth and magnificence were then unrivalled in the East, and have been celebrated by the Persian writers in many a romantic volume. The Arabian found him in his hall of audience, surrounded by the officers, musicians, and ladies of the court, of whom there were twelve thousand, "every one equal to the moon in beauty." A huge crown, wrought of the most costly jewels and pearls, and compared by Tabiri to an Eastern measure containing six bushels of wheat, was suspended over the throne by a golden chain from the roof; the weight being too much for the royal brows to support. It was covered with a veil, which was never removed except on state occasions.

Khoosroo listened with indifference to the invitations of the suppliant prince. "Thy land," said he, "is distant and barren. Its only productions are sheep and camels; these we want not; nor can they tempt the Persians to so fruitless an enterprise." He then ordered Seiph a thousand pieces of gold, and a robe of curious workmanship. The wily Arabian immediately threw the gold to the slaves and the crowds in the streets. "Of what use," said he to the astonished monarch, "is the wealth or the jewelry of Persia to me? The hills of my own country are of gold, and its dust silver." This appeal to the avarice rather than the sympathies of the Persian had better success. Khoosroo ordered a levy of all the prisoners and condemned criminals within his dominions, to the amount of 3600 men. "If they conquer these regions," said he, "it will add to my kingdom; if they perish, they but suffer the just punishment of their crimes." With these auxiliaries Seiph returned to Arabia; a battle was fought near Aden, where Masruk fell by an arrow from the hand of Wehraz, a Persian nobleman, who commanded the expedition. The victory diffused universal joy among the Arabs. The successful general took possession of Sanaa, where he put all the Abyssinians to death, and planted his master's standard on its walls; having thrown down one of the gates, rather than lower the proud banner of Khoosroo.

Seiph was made viceroy in name of the Persian king, and commanded to pay an annual tribute. His cruelties to the Abyssinians occasioned a conspiracy, and after a reign of four years he was waylaid and stabbed by a slave while hunting in the neighbourhood of his own capital. Wehraz inflicted a cruel retaliation, by putting to death every man with the dark skin and crisp hair of Ethiopia, to the number of about three thousand. From this time until it was subdued by the lieutenants of Mohammed, Yemen was governed by Persian satraps, under the title of emirs; the last of whom, Badsan, submitted to the faith of the Prophet. Thus, in less than a century, the arms of Khoosroo supplanted the Abyssinian power in Arabia. Had Abraha succeeded in demolishing the Kaaba, and establishing the Christian worship on its ruins, Arabia might have acknowledged the apostleship of St. Peter instead of the impostor of Mecca, and quietly submitted to the doctrines of the Cross, without undergoing the shock of a revolution, which has changed the civil and religious state of half the world.*

II. The kingdom of Hira, in Irak, was founded by a part of the dispersed clans which the flood of El Arem had compelled to abandon Yemen. These emigrants first settled on the borders of Hejaz, where they all remained till the death of Amru Mazikia. The scanty produce of that country being inadequate to the maintenance of so great an influx of strangers, another dispersion became necessary. The tribe of Tai took up their residence in Nejed, in the district of Mount Salma; that of Khozaa continued at Mecca, where they succeeded for a time in wresting from the Ishmaelites the superintendence of the temple and the principality of the city. Malec, with the tribe of Azd and Khodai, a powerful colony, settled in Bahrein and Yemama. The throne of the Arsa-

* Schulten's Hist. Joctanidarum. Abulfed. Hist. Gen.

cidæ still subsisted in Persia, but in circumstances of such feebleness and disunion as invited the wandering Arabs, about the beginning of the third century, to take possession of Irak, which they found without any regular government or means of defence. They made themselves masters, for a time, of the whole territory lying between the passes of Hulwan and the Tigris. Malec fixed the seat of his new kingdom at Anbar, on the Euphrates, where certain of his countrymen, known by the name of Armenians, were already settled, who had been carried from Arabia among the captives of Nebuchadnezzar. The capital was afterward transferred to Hira, a city lower down the river, by Amru, the third prince of this dynasty, with whom the throne passed by marriage from the descendants of Cahlan to the Lakhmians, another branch of the royal house of Saba in Yemen. The following list of the kings of Hira is furnished by Pococke and Hamza.*

* Abulfeda and Hamza assign this kingdom a duration of 622 years, evidently reckoning from the dispersion of El Arem; Tabiri, 489 years, five months; exclusive of three reigns, the length of which is not expressed. De Sacy fixes Malec's reign A. D. 210; but he admits that it may be more ancient. "On pouvroit, si l'on vouloit, faire remonter un peu plus haut, l'établissement de Malec dans l'Irak. Je regrette de ne pouvoir employer ici l'ouvrage de Hamza." The parts of Hamza relative to the kings of Hira and Gassan, which he had not seen, have since been published by Rasmussen, late Prof. of Orient. Lit. at Copenhagen, in his Hist. Præcip. Arab. Reg. ante Islam. 1817. According to this chronology, the kingdom of Hira must have commenced about A. D. 12. It is evident that some of the early princes settled in Bahrein have been omitted; but the Arabs fill up the chasm, as usual, by making Amru reign 118, and Amriolkais 114 years.—*Price's Essay*, chap. iv. vii. "At this time probably happened the migration of those colonies which were led into Mesopotamia by the chiefs Becr, Modar, and Rabiah; from whom three provinces in that country are still named,—Diyar Becr, Diyar Modar, and Diyar Rabiah."—*Golii Nott. ad Alfragan*, p. 232

Kings of Hira,—Reigned A. D. 210–634.

A. D. 210, Malec.	A. D. 504, Abujafar.
230, Jodaimah (or Khozzeimah).	507, Amriolkais III.
DESCENDANTS OF LAKHM.	520, Mondar III.
268, Amru I.	523, Hareth.
301, Amriolkais I.	563, Mondar III. (restored).
334, Amru II.	564, Amru III
369, Aus.	576, Kabus.
374, Amriolkais II.	584, Mondar IV.
400, Nooman I.	588, Nooman III.
430, Mondar I.	611, Ayas.
473, Aswad.	617, Zadijah.
493, Mondar II.	634, Mondar V.
500, Nooman II.	

The history of these kings presents little that is worthy of particular notice. War was their incessant occupation; and there is a tradition that none of them except Kabus died within their own territories, all the rest having perished either in military expeditions or hunting excursions. Malec, on assuming the functions of sovereignty, proceeded to establish the absurdities of the Arabian idolatry throughout his dominions. He was slain accidentally with an arrow by a person named Soleimah, while wandering in disguise from his palace to observe what was passing in the town. Jodaimah, on whom was bestowed the title of Al Abrash or the Leper, was a brave and judicious prince; he succeeded in subjugating to his power the whole of the Arab chiefs settled in Irak. His authority was acknowledged in Bahrein, and even extended to some parts of Hejaz and Yemen. He introduced regular discipline among his troops, which gave him great advantages over the desultory tactics of his adversaries. It is remarked that he employed lamps in his nocturnal marches, and was the first that used the balista, a military engine (perhaps the crossbow) for throwing missiles. In all his expeditions he carried with him two images or idols; from the one he supplicated health, from the other victory. He was attacked and defeated by Asaad-Abucarb and Hassan, kings of Yemen; but the disorders that broke out among the ranks of the invading army saved the kingdom of Hira. Amru I. was the son of Rakash, the sister of Jodaimah, and the handsome Addi, chief of the Beni-Lakhm, who filled the office of the king's cupbearer; and in his tribe the sovereign power now became hereditary. This prince threw off for a long time his dependence on Persia, until Shapoor, the son of Ardisheer Babigan, entered Irak with a numerous force, and subdued the greater part of the country between the Tigris and the Euphrates.

A debt of revenge had descended to Amru in the murder of Jodaimah, who was entrapped by the treachery of Zabba, an Arabian princess, in Mesopotamia, and ordered to be bled to death by opening the veins of his arms. This cruel deed demanded an atonement equally severe; and it was effected by the following stratagem:—Kosair, a confidential servant, was despatched in the disguise of a merchant with a large caravan to Khadr, a strongly-fortified city, where Zabba had her palace. Some assert that, to excite pity, he had, like another Zopyrus. disfigured and mutilated himself by cutting off his nose. Pretending to have brought rich merchandise, which he wished to submit to the princess, the gates were without hesitation thrown open. The cargo consisted of 2000 large sacks of hair-cloth, each of which concealed two armed men, who, on a given signal, surrounded the royal residence, putting all to slaughter without resistance. Zabba fell by the hand of Kosair, or by swallowing a deadly poison which she kept enclosed in a ring. Amru annexed her territory to that of Irak, and transmitted the government, which he had thus consolidated, without interruption to his posterity.

Nooman I. signalized himself by his conquests in Syria. The immense spoils he collected increased the wealth of his kingdom, and enabled him to adorn his capital with gardens, vineyards, groves, and hunting parks, not inferior to those of Mareb. The Euphrates was covered with his boats and pleasure-barges; and his preserves were richly stocked with gazelles and other animals of chase. To the care of this prince the Persian monarch, Yezdijird, intrusted the education of his only surviving child and successor, the celebrated Baharam-Gour (the Varanes of Roman history), who is reported to have won back, in a dispute with the usurper Khoosroo, his father's crown, by carrying it off from his less daring competitor when placed between two furious lions. It was chiefly for the accommodation of his royal pupil that Nooman erected those magnificent buildings or castles called Khavarnak,—the Palace of Delights,—and Sadir, reckoned the most charming and salubrious residence in all Irak. The imagination of the Arabs has described them as altogether unrivalled in elegance and splendour; but the unlucky architect, Sennamar, having incautiously admitted that he had not expended the utmost of his skill on those boasted structures, was precipitated by order of Nooman from one of the loftiest towers. Hence the proverbial expression applied to a person ungratefully used or inadequately remunerated for his labours, that "he has met with the reward of Sennamar."*

Nooman is said to have become a convert to Christianity, when he abdicated the throne, and, like another Charles V., retired from the cares and functions of royalty, to moralize on the vanity and evanescence of all sublunary grandeur. Conducting his courtiers to the top of Khavarnak, he pointed to the watered gardens and palm-groves, the pastures

* "The Arabs," says D'Herbelot, "reckoned this palace one of the wonders of the world. A single stone fastened the whole structure; the colour of the walls varied frequently in a day."—*Biblioth. Orient.*

covered with flocks, the river with its crowded barges, and the city with its busy population. "Of what worth," he exclaimed, "are those fleeting possessions to me! to-day they are mine; to-morrow they belong to another!"

On his accession to the throne, Mondar I. led an army of 40,000 men into Persia to assist his youthful companion, Baharam-Gour, in recovering the crown which had been factiously conferred on another. His reign and those of his immediate successors are barren of incident. The energy and admirable political regulations of the Persian monarchs, Kobad and Nooshirwan, gave little cause for mutiny or discontent in their tributary provinces. In India and Arabia, from the Mediterranean to the Caspian, and from the Euxine to the Jaxartes, the imperial sway of the latter was acknowledged; and it was not till the sceptre fell into the hands of his unworthy successor, Hormuz III., that the Arabs refused to pay tribute or obedience.

Mondar II. had proved a valuable ally to Kobad in his wars against the Roman emperor Anastasius. Mondar III. was defeated and taken captive by the tribe of Becr, who raised Hareth to the throne; the Persian monarch being too much occupied in suppressing the religious innovations of the impostor Mazdak in his own dominions, to protect his vassal or attend to the affairs of Hira. In 531, Mondar joined Kobad in his successful invasion of the Roman territories, and marched to his assistance with an army of 150,000 men. He advised him, as we learn from Procopius, to alter his plan of hostilities; and, instead of carrying on the war in Mesopotamia, to penetrate directly into Syria, where there were no fortified cities, and lay siege to Antioch, one of the richest and worst-defended places in the Eastern Empire. His counsel was accepted; and an expedition, which he himself conducted through the desert, was despatched against the famous Belisarius, whom he soon completely routed near the Euphrates. Mondar had the reputation of a bold and experienced soldier. "For fifty years," says Procopius, "he had harassed the Romans, from Egypt to Mesopotamia; pillaging their country, burning their cities, and making innumerable slaves, whom he killed or sold for large sums of money. He made his inroads so suddenly, that he was off before any general was apprized of the attack, or could pursue him with advantage. He captured many Roman officers, and exacted large sums for their ransom. In short, he was the sharpest enemy the Romans had; and the defeats they had sustained at his hand induced the Emperor Justinian to associate the Hamyarites and Ethiopians against the Persians."*

The remaining kings of Hira demand few remarks. Nooman III., from being a tyrant and an idolater, is alleged to have become a convert to the doctrines of the Cross, by witnessing the devoted friendship of a Christian Arab who had pledged himself, as Pythias did to Damon, to undergo the punishment intended to be inflicted on his friend, should the latter fail to return at the time appointed. Struck with this heroic magnanimity, he pardoned both the criminal and his surety, and embraced a religion capable of inspiring such noble sentiments. He was slain in battle by Khoosroo-Purvees, who overran Syria and Palestine with his sudden victories, took Jerusalem, and plundered the great church; carrying off with him to Persia, among other sacrilegious spoils, the true cross enclosed in a case of gold. It was when Mondar V. occupied the throne that the kingdom of Hira was invaded and subdued by the lieutenant of Mohammed (A. D. 633), when its history becomes incorporated with that of its Moslem conquerors.

III. The other colony of the dispersed Arabs, of

* Procop. De Bello Persico, cap. iii.

which we have already spoken, migrated northwards into the territory of Damascus, where they founded a dynasty of kings called the Gassanites (probably the Cassanitæ of Ptolemy), who derived their name from Gassan, a fertile valley with a well, whose waters they found so pleasant and convenient as to induce them to settle on the spot. Several small principalities existed in those districts before their arrival; the chief of which were the tribe of Salih, who had embraced Christianity, in consequence of which the Roman emperor invested them with the government of all the Syrian Arabs. These the emigrants (the tribes of Aus and Khazraj) expelled, slew most of their petty kings, and established their own sovereignty over the vanquished territories.* The duration of this monarchy, which comprehended thirty-two kings, includes a period of little more than 400 years according to some; Abulfeda computes it more exactly at 616; while Nuvairi and others extend the list to thirty-seven successions. The following table is supplied by Hamza:—

Kings of Gassan,—Reigned A.D. 37-636.

	Reigned.		
	Years.	Months.	A.D.
Jafnah (or Haneifah)	45	3	37
Amru I.	5	—	87
Thalabah	17	—	104
Hareth I.	20	—	124
Jabalah I.	10	—	134
Hareth II.	10	—	144
Mundar-Al-Akbar (the Great)	3	—	147
Nooman I.	15	6	162

* The cause of quarrel was the tribute which the Salihites wished to impose on the colonists, in recompense for the privileges they had granted them. Wearied with the importunities of the tax-gatherer, the intepid Gadza presented a golden sword by way of pledge or substitute for the tribute; and while the obnoxious officer was in the act of seizing it, he stabbed him to the heart. Hence the proverb, the Gift of Gadza.—*Rasmussen*, ***Hist** Præcip. Arab. Reg.—Notæ.*

	Reigned Years.	Months.	A.D
Mondar II.	13	—	175
Jabalah II.	34	—	209
Ayham I.	3	—	212
Amru II.	26	—	238
Jafnah II.	30	—	268
Nooman II.	1	—	269
Nooman III.	27	—	296
Jabalah III.	16	—	312
Nooman IV.	22	5	334
Hareth III.	37	—	371
Nooman V.	18	—	389
Mondar III.	19	—	408
Amru III.	33	—	441
Hajar	12	—	453
Hareth IV.	16	—	479
Jabalah IV.	17	1	496
Hareth V.	21	5	517
Nooman VI.	37	3	554
Aiham II.	27	2	581
Mondar IV.	13	—	594
Sharahil	15	3	619
Amru IV.	10	2	629
Jabalah V.	4	—	633
Jabalah VI.	3	—	636

The Gassanite kings, from their being of the Christian religion and viceroys of the Roman emperors, have been but slightly noticed by the writers of Arabian history, These annalists speak of the number of churches or monasteries erected by them, and of the wars in which they were frequently engaged with the sovereigns of Persia and Hira. Hareth, called Aretas by the Greeks and Latins, appears to have been a general appellative among the petty chiefs of the Arab tribes in various parts of the country Long before the origin of the Gassanites, several princes of this name, some of them very powerful, are mentioned in Josephus and the Maccabees as seated on the Syrian frontier, in the Desert and Stony Arabia. We learn from St. Paul, that in his time (A. D. 34) Damascus was ruled by an Arabian king called Aretas. In the reign of Hormuz II

(A. D. 310) a tribute being demanded by the Persians from the Syrian Arabs of Gassan, the latter, trusting to the protection of their Roman allies, not only refused to comply with the exaction but even ravaged the fertile valleys of Irak; and having waylaid Hormuz in the desert, put him and his whole attendants to the sword. These indignities were dreadfully retaliated by his son Shapoor, surnamed Zoolactaf or Lord of the Shoulders, from the punishment he inflicted on the shoulders of his captives by causing them to be pierced, and dislocated with a cord passed through them. Pursuing his career of slaughter, he crossed the desert with great rapidity to Medina, massacring every Arab he met with, and filling up all the wells on his march. From Hejaz he turned northward, and appeared before Aleppo, extending in every direction the dire effects of his vengeance. The kingdom of Gassan, like that of Hira, was extinguished by the swords of the first Mohammedan conquerors.

Besides the three principal dynasties, of which we have given some account, there was a variety of other smaller states which had princes of their own, who acknowledged no dependence on any superior. The tribes of Kenda, Maad, and Kelab had several kings distinguished for the wisdom of their government and the lustre of their fame; though no other monuments of their reign have been recorded than the civil feuds in which they were almost incessantly engaged. Except in the two instances already mentioned, no other female appears in the list of Arabian sovereigns. That such occurrences, however, were not unfrequent, we learn from different sources; and it was a reproach of their enemies, that a nation of warriors allowed themselves to be governed by women. The province of Yemama is said to have been so called from a queen of that name; and if we may believe the Chaldean interpreter of Job, a royal virago named Lilith, headed the Sabæan brigands who despoiled that wealthy patriarch. Ecclesiastical historians repeatedly mention Moavia as among the female rulers of the Arabs. And we learn from Ammianus Marcellinus, that after the defeat and death of the Emperor Valerius, when the Goths laid siege to Constantinople, this princess sent a body of her best troops to assist the Romans; and it was principally by their bravery that the barbarians were forced to retire.*

The Nabathæan or Ishmaelite Arabs, under a race of native princes, long preserved a distinct name as a nation; asserting their independence alike against the hosts of Egypt and Ethiopia, of the Jews, the Assyrians, the Greeks, and the Romans,—all of whom had successively assailed their territories. Diodorus Siculus states, that Sesostris, who reigned 1300 years before the Christian era, harassed by their incessant depredations, was compelled to draw a line of defence across the isthmus of Suez, from Heliopolis to Pelusium, to secure his kingdom against their incursions. It was extremely difficult, adds this author, either to attack or subdue them, because they retired to their deserts; where, if an enemy ventured to follow, he was sure to perish of thirst and fatigue, for their wells were only known to themselves.†

The dukes of Edom were famous long before there reigned any king over the Israelites (Gen. xxxvi.), and they refused Moses a passage through their territories to the Land of Canaan (Num. xx.)

* Saraceni mulieres aiunt in eos regnare.—(Exposit. Tot. Mund. Hudson's Geog. Min. tom. iii.) Multos annos occupavit fæmina regnum Sabæ, et post eam fæminæ usque ad tempus Solomonis, filii Davidis. Poc. Specim. p. 85. Ammian. Mar cell. lib. xxxi. Philostorg. Hist. Eccles. lib. iii. cap. 4.

† Diodor. Sic. lib. ii. cap. 4. Ammian. Marcell. lib. xiv cap. 4.

They were conquered by David, who planted Hebrew garrisons at Elath and Ezion-gaber; and probably commenced the trade of Ophir, which was afterward carried to its height by Solomon and Hiram. For about a hundred years they continued in submission to the Jewish sovereigns; but in the reign of Jehoram they shook off the yoke, and maintained their independence until they were again subdued by Uzziah, after an interval of eighty years. More than two centuries later they were subject to Nebuchadnezzar, and assisted that monarch when he besieged Jerusalem.

In the time of the Assyrian and Babylonian empires, which put a period to the kingdoms of Israel and Judah, these wild freebooters remained either entirely their own masters, or acknowledged a temporary alliance with their enemies. When the great Eastern monarchy fell before the invasion of the Medes and Persians, under Cyrus, Cambyses, and Darius, the conquerors found it necessary to keep up a friendly understanding with the tribes of the desert, in order to obtain a passage through their territories into Egypt; to the subjugation of which their assistance materially contributed, by supplying the invading armies with water and provisions on their march. Herodotus observes, that on this account the Arabs were exempted from paying tribute; while the neighbouring provinces, Syria, Palestine, Phenicia, and Cyprus, forming the fifth satrapy of the Persian empire, were taxed at 350 talents.

Belesis and Sennacherib are called kings of Arabia; but it is evident they had assumed the title without possessing the country. Xenophon alleges that Cyrus reduced it to subjection. It appears, however, that his conquests were limited to the few tribes seated on the borders of Phrygia and Cappadocia, which he encountered on his march from

Sardis to Babylon.* The project of Alexander the Great, after he had overthrown the Persian empire, to add Arabia to his dominions, was frustrated by a premature death. That prince, as we learn from Arrian and Strabo, intended to fix his royal residence there; and it was with this view that the fleet of Nearchus, the Columbus of antiquity, was ordered to make a survey of the whole peninsula.

Antigonus sent an army of 4000 foot and 600 horse to chastise the Nabathæans for the ravages they had committed, and their refusal to allow him to collect the bitumen of the Lake Asphaltites; but they were taken by surprise, and almost entirely cut to pieces. None of Alexander's successors appear to have made any efforts to extend their authority beyond the frontier districts of Arabia. When the Macedonian empire was partitioned into four kingdoms (B. C. 301), this country was indeed included as a province in that which fell to the inheritance of Ptolemy; but the name is evidently applied merely to the regions that bordered on Egypt and Palestine. Ptolemy Euergetes had made himself master of the Arabian and Ethiopian coasts of the Red Sea; but he penetrated no farther into the country.†

From about the year 220 B. C., to the Christian era, several of the Arab kings distinguished themselves in the wars in which the Jews were engaged, sometimes joining the Syrians, and sometimes the

* Herodot. lib. i. iii. Prideaux's Connex. vol. i. p. 1, 47, Xen. Cyropæd. lib. viii. p. 515, &c.

† The nations conquered by Ptolemy are pompously recorded in the famous Adulitic Inscription, which mentions his having subdued "the whole coast from Leuké Komé to Sabæa; and his being the first to conceive the design, and carry it into execution." But for this curious monument, which was preserved by Cosmas Indicopleustes, in his Topographia Christiana, the victories of Ptolemy might have been buried in oblivion. They paved the way for the Abyssinian power in Arabia.—*Vincent's Periplus*, vol. i. Append. ii. *Valentia's Travels*, vol. iii chap. 5.

Egyptians. Antiochus the Great reduced part of the northern tribes to submission, and his son Hyrcanus was occupied several years in chastising their incursions and depredations. At that period (B. C. 170) the Nabathæans were ruled by a prince named Aretas (Hareth), whose dominions extended to the confines of Palestine, and included part of the land of the Ammonites. Having made peace with the Jews, they permitted Judas Maccabeus and his brother Jonathan to pass through their territories; but, notwithstanding the amity subsisting between them, they could not resist the temptation of pillaging even their friends when an opportunity offered; and, accordingly, they fell upon a detachment of their forces while on their march, seized their carriages, and plundered the baggage. Zabdiel, another of their princes, afforded protection to Alexander, king of Syria, when defeated by Ptolemy Philomater (B. C. 146); but the influence of money induced him to violate the laws of hospitality, by delivering up the royal fugitive into the hands of Tryphon. Aristobulus, according to Prideaux, forced a tribe of the Ishmaelites to become proselytes to the Mosaical religion. Josephus mentions an Arab prince, whom he calls Obodas (Abd-Waad), who defeated the Jews (B. C. 92), by drawing them into an ambuscade, where their king and the greater part of his army were cut to pieces. We learn from the same author that Aretas, ruler of Arabia Petræa, overthrew Antiochus Dionysius, the sovereign of Damascus; and some years afterward, having advanced to Jerusalem with an army of 50,000 men, he defeated Aristobulus; but returned home on finding that the Romans had espoused the interests of that prince.*

The repeated inroads of the Arabs into Syria pro-

* Josephus, Antiq. lib. xiii. cap. 1, 9, 19, 23. De Bell Jud. lib. i. cap. 4, 5. 1 Maccab. chap. v. ver. 24–36; chap. xi ver. 15–18, &c. Prideaux's Connex. vol. i. p. 11; vol. ii. p. 188

voked the wrath of the Cesars, whose empire at this period extended as far as the Euphrates. Lucullus, Pompey, Scaurus, Gabinius, and Marcellinus, all proconsuls of Syria in succession, undertook expeditions against them, under their kings, Aretas, Malchus (Malek), and Obodas; without, however, gaining any other advantage than the payment of tribute, or a temporary cessation of hostilities. Antony and Herod defeated Malek in a general action, and compelled him to render an annual impost to Cleopatra for certain portions of his territories bordering on Egypt; but the tax was remitted on the death of the Jewish prince. Plutarch mentions Agbar, an Arabian emir, who misled Crassus to his destruction, and assisted Pompey in his expedition against Petra.* Augustus pretended to the right of imposing a new king on the Arabs; but they elected a sovereign of their own, who assumed the name of Aretas, and continued, as his predecessor Obodas had done, on friendly terms with the Romans until his death, about forty years after the Christian era.†

It was in the reign of Augustus that Ælius Gallus, prefect of Egypt, undertook his famous expedition into Yemen, being the only Roman general who penetrated into that country. His force consisted of 80 ships of war, 130 transports, 10,000 Roman infantry, 500 Jews, and 1000 Nabathæans. The emperor had heard of the extreme wealth of the Sabæans; and the object of this formidable armament was either to conquer or conciliate them. The success of the enterprise was not equal to its mag-

* Plut. in Pomp. Dion Cassius, lib. xlviii. p. 234. Appian. De Bell. Civil. Hirtius speaks of Malek as one of the allied kings to whom Julius Cæsar sent for cavalry. De Bell. Alexand.

† This was the Aretas mentioned by St. Paul, whose vengeance he eluded, being let down from a window in a basket.—2 Cor. xi. 32.

nitude. Syllæus, an officer of Obodas, who had undertaken to guide the invaders, deceitfully conducted the fleet to Leuké Komé (Moilah), through the most dangerous part of the Red Sea, so that many of the ships were lost among the rocks and shelves. New treacheries and disasters awaited the troops by land; disease, fatigue, and famine daily reduced their numbers. The prefect, however, continued his march without much opposition. In eighty days after leaving Moilah, and passing through the desert tracts of Medina and Mecca, he reached Negrana (Nejeran), which he took by assault,—the inhabitants with their chief having fled to the mountains. A march of six days more brought him to the cities called Asca and Athrulla, both of which surrendered after a bloody battle, wherein 10,000 of the Arabs were slain. These successes enabled him to advance to Marsyaba or Mariaba, which is represented as the capital of the Rhamin-ites, and the seat of Alasar, the sovereign of that country. Here, according to Strabo, the expedition terminated; six months having elapsed since its outset. The distresses of the soldiers rendered a speedy retreat necessary. In sixty days Gallus reached the country of Obodas, and crossing the gulf at Myos Hormus, he arrived at Alexandria with the shattered remains of his army. The traitor Syllæus paid with his life the forfeit of his perfidy; he was sent to Rome, and beheaded by order of the emperor. In his report to the senate, the general related what he had heard of these celebrated nations of Arabia Felix,—that the Homerites were the most numerous,—that the Minæans were rich in palm-groves, flocks, and fertile fields,—and that the Sabæans were renowned above all others for their odorous woods, their well-watered gardens, their mines of gold, their abundant stores of wax and honey, myrrh and frankincense.

There is much obscurity and contradiction in the

different accounts of this Roman expedition. Were it possible to identify Mariaba with Mareb, the ancient metropolis of the Hamyarites, the cause and date of the mysterious flood of El Arem would admit of an easy solution; but the confused statements on the subject have left this important point merely a matter of plausible conjecture. Strabo relates that the Romans, after lying before the place for six days, were compelled to raise the siege for want of water; a necessity which could not have existed in the vicinity of an immense reservoir. Pliny, on the contrary, includes Mariaba, which he describes as the capital of the Calingii, and six miles in circumference, among the cities that were wasted or destroyed; and he calls the town where the expedition terminated Caripeta. Gossellin and De Sacy rather think that Mariaba must have been Mecca (Macoraba); but this place did not belong to the Sabæans, and the distance from Moilah (about 620 miles) is obviously too short to occupy so tedious a march as six months. Besides, had such been the fact, it is scarcely credible that the Arabs would have passed in total silence so memorable an insult to their sacred city.

Dr. Vincent is of opinion that the Mariaba destroyed by the legions of Gallus was the capital of the Minæans in Tehama, and not of the Sabæans; the distance of the latter from Moilah (above 1000 miles) he considers as too great to be accomplished in a retreat of sixty days. This objection, however, is far from being conclusive. No city of that name is known to have existed in Tehama. Nor is there any insurmountable objection in the remote situation of Mareb; for we learn from Burckhardt, that the same journey is still performed by the caravans in sixty-two days of "slow travelling;" and we can hardly suppose that a retreating army, which most likely took the same route, and whose movements were accelerated by famine and an enemy hanging on their rear, would require more time on the march than is occupied by the pilgrims of the present day. The hypothesis of D'Anville, that identifies Mariaba with Mareb, is therefore as probable as any other; and the catastrophe arising from the bursting of the mound was, perhaps, after all, the work of the Roman army. The two events certainly accord very nearly in point of time; and as Gallus was informed that the city lay within two days' march of the region that produced aromatics. there is a strong presumption, notwithstanding the confusion of names and the silence of the Arabs, whose national pride might prompt them to ascribe this disaster to the operation of local causes rather than the invasion of an enemy, that the famous deluge, which desolated the capital of Yemen and dispersed so many tribes, must be referred to this adventurous inroad of the Egyptian prefect.*

When Titus laid siege to Jerusalem, a body of Arabian auxiliaries, as we learn from Tacitus, accompanied his army; but till the days of Trajan, whose fleets ravaged the coasts of the Red Sea, their country is little noticed in Roman history. In the year 106, his lieutenant, Cornelius Palma, governor of Syria, reduced Petræa to the form of a province, under the name of Palestina Tertia or Salutaris; but the fluctuating power of the empire was unable to retain it in a state of absolute dependence. The flatterers of Trajan, especially Lucian, Dion, and Eutropius, have numbered among his other victories the subjugation of all Arabia, and part of India; and coins were actually struck in commemoration of these exploits;—monuments which prove nothing but the fulsome adulation of their authors, or the excessive vanity of the Romans. With the exception of Severus, who conducted a numerous

* Strabo, lib. xvi. Pliny, lib. vi. cap. 28. Vincent's Periplus vol. ii. Horace alludes to the failure of this expedition.—*Carm.* lib. i. ode 35.

but unsuccessful army against certain tribes (A. D. 194), to chastise them for assisting his rival Piscennius Niger, none of the emperors appear to have disturbed the repose of their desert, till Aurelian, who, in A. D. 273, vanquished the celebrated Queen Zenobia, and made himself master of her capital, Palmyra. A number of Arabs were among the captives of various nations who graced the splendid triumph of that warlike prince on his final return to Rome.*

The apostate Julian paid that people a yearly sum, to maintain a body of troops in readiness for his service; and when their deputies complained that this stipend had been discontinued, he imprudently remarked that a military chief should pay with steel, and not with gold. This haughty neglect they immediately resented by joining his enemies, the Persians. The love of plunder, however, attracted several tribes to the imperial standard, when the same emperor conducted his expedition (A. D. 362) against the warlike Shapoor, who had insulted the majesty of the Cesars by threatening to exterminate with his fiery scimitar the name of Romans from the earth. It was in this memorable campaign that the soldiers of Julian besieged and destroyed Anbar, the residence of the kings of Hira, which Ammianus describes as a city, large, populous, and well fortified with a double wall; almost encompassed by a branch of the Euphrates, and defended by the valour of a numerous garrison. After being battered for two days, by two prodigious engines resembling moving turrets, the place was reduced to ashes. Two thousand five hundred persons, the feeble remnant of a flourishing people, were permitted to retire. The magazines of corn, arms, and ammunition were partly distributed among the troops, and partly reserved for the public service. The rich furniture and superfluous stores were destroyed by fire, or

* Dion Cass. ut sup. p. 777. Lucian, Philopatris, Eutrop. in Traj. et Sever Gibb. Rom. Emp. chap. xi.

thrown into the neighbouring stream. Down to the Mohammedan era, the Arabs continued to commit their usual depredations on the frontiers of the Eastern Empire. Theodosius had tried in vain to check their irruptions (A. D. 452); but Marcian forced them to sue for peace, which he granted on terms highly advantageous to the Romans.

The perpetual independence of the Arabs has been the theme of praise among strangers and natives, and the subject of controversy between the skeptic and the Christian. This singular fact is at once the fulfilment of a prophecy, and the effect of their local position. The obvious causes of their freedom are to be found in their character and their country. Surrounded with inhospitable deserts, they could easily elude the vengeance of their enemies, by retiring within those natural barriers of ocks and sands which bade defiance to their pursuers. In this manner they preserved their liberties, because it was impossible to penetrate their retreats. The exceptions to the perpetuity of Arabian independence, which Gibbon alleges can neither be dissembled nor eluded, do not in the slightest degree invalidate the predictions of Scripture as applied to the posterity of Ishmael. As a nation they have never been conquered. Their subjugations have been partial and temporary; they escaped the yoke of the most powerful monarchies of antiquity; and in modern times the precarious jurisdiction of Turkey and Egypt scarcely extends beyond their frontiers. The sneer of Gibbon is thus refuted by the facts of history; and though the evidence of Christianity rests not on the habits or independence of this remarkable people, we cannot join that learned historian in his blush at the "nameless doctor" who has made these circumstances a formal demonstration of its truth. The "wild man" still spurns the chains of a foreign conqueror,—still waylays the traveller by the fountain; and maintains himself, as in the days of old, by violence and plunder, sweeping his troop of fierce bandits across the path of the merchant and the pilgrim.*

The same peculiarities that secured the independence of these roving hordes against the disciplined legions of the East and the West, prevented them from acquiring influence, or extending their conquests beyond their own territories. While the Babylonians, Assyrians, Medes, and Persians, on the one side, and the Egyptians, Greeks, and Romans, on the other, rose in succession to the proud eminence of imperial grandeur, and spread their victories over the greater portion of the known world; scarcely a gleam of splendour shines on the long dynasties of the Tobbaas, the Hareths, and the Mondars of Arabian history. Nor is it difficult to explain the cause. The parcelling of the nation into independent tribes impaired their common strength; no necessity ever summoned them to combine for their mutual defence; no motives of external advantages could prevail with them to suspend their domestic feuds; and no leader till Mohammed arose seems to have possessed the genius or address to concentrate their impetuous energies, with a view to national aggrandizement.

Connected with this part of our history, there remains to be noticed one of the most singular spots in all Arabia,—perhaps in the Eastern World,—Petra, the ancient capital of Idumæa, which has been but recently brought to light, after being for a series of ages as effectually hidden in its solitude

* The historian of the Decline and Fall of the Rom. Emp., in a note, chap. l., says, "The 'non ante devictis Sabeæ regibus, and the 'intacti Arabum thesauri,' of Horace (Odes i. 29 and iii. 24), attest the virgin purity of Arabia." Yet, despite of these classical facts, and of his own quotations, this learned skeptic, with his usual inconsistency when treating of religious subjects, labours to prove, almost in the same page, that the perpetual independence of the Arabs is an unfounded boast.

from the knowledge of Europeans as the palace of Sheddad, or the fabled paradise of Irem. This city appears to have been coeval with the birth of commerce; and there is indubitable evidence that it was a flourishing emporium seventeen centuries before the Christian era. It was the point to which all the trade of Northern Arabia originally tended; and where the first merchants of the earth stored the precious commodities of the East. It formed the great *entrepôt* between Palestine, Syria, and Egypt, and there is little doubt that the company of Ishmaelites with their camels, bearing spicery, balm, and myrrh, to whom Joseph was sold by his brethren, were the regular caravans that visited the markets of Petra. The famous soothsayer Balaam was a native of this place, whose inhabitants were then renowned for their learning, their oracular temple, and their skill in augury.*

After Antigonus had recovered Syria from Ptolemy, he sent two successive detachments, under his general Athenæus and his son Demetrius, to take Petra by storm; but both expeditions failed. It was besieged by Severus, the Roman proconsul, who found the place to be impregnable. By the advice of Antipater, he agreed to take a sum of money and raise the siege. Lucullus and Pompey had no better success. The latter was obliged to come to terms with Aretas; and the former could only obtain a temporary truce, notwithstanding the insinuations of Plutarch that he had subdued the whole nation of the Arabs. Trajan, who put an end to the dynasty of its ancient kings, invested this capital with a numerous army; but, from its strong position, and the gallant defence of the garrison, he found its reduction impossible. In one of the assaults, which he headed in person, he narrowly escaped being slain; his horse was wounded, and a soldier killed

* Good's Translat. of Job, note, p. 37. Numbers xxii. 5.

by his side; for the Arabs, notwithstanding his disguise, discovered him by his gray hairs and his majestic mien. The Romans were compelled to abandon the siege; a repulse which the historians of the times ascribe to the violent storms of wind and hail, the dreadful flashes of lightning, and the swarms of flies that infested the camp of the besiegers. Under several of the later emperors, Petra appears to have continued the seat of wealth and commerce. Strabo did not visit it himself, but he describes it from the account of his friend Athenodorus the philosopher, who spoke with great admiration of the civilized manners of its inhabitants, of the crowds of Roman and foreign merchants, and of the excellent government of its kings. The city, he adds, was surrounded with precipitous cliffs; but rich in gardens, and supplied with an abundant spring,* which gave it a distinction from all the rocks in the vicinity, and rendered it the most important fortress in the desert. Pliny describes it more correctly, as a town nearly two miles in extent, with a river running through the midst of it, and situated in a vale enclosed with steep mountains, by which all approach to it was cut off.

With the decline and fall of the Roman power in the East the name of Petra almost vanishes from the page of history. About the period of the Crusades, it was held in such esteem by the sultans of Egypt, on account of its great strength, that they made it the depository of their choicest treasures; and, in course of these religious wars, its possession was strenuously contested by the Turks and Christians, who regarded it as the key that opened the gates of Palestine. From that time it was known only as the seat of a Latin bishop. Its once crowded marts ceased to be the emporium of nations. The obscurity of nearly a thousand years covered its

* Called Thamud by Edrisi (Geog. Nubian.); hence the old tribe of Thamudeni probably took their name.

ruins. The very place where it stood became a subject of controversy. That it was the Rekem of Moses and Joshua, Rakeme of Josephus, and the Hajr of the Arabs, all synonymous with the Petra of the Greeks (a rock), was generally admitted; but until the present century its situation was unknown, or mistaken for the town of Kerek, near the border of the Dead Sea, which was also a strong fortress of the Nabathæans at the time of their first acquaintance with the Greeks and Romans. That Kerek was then the capital of Petræa appears probable from Strabo's description, who says, that when besieged by Demetrius the Arabs placed their old men, women, and children on a certain rock (ἐπί τινὸς πέτρας), steep, without walls, admitting only of one access to the summit, and situated 300 stadia from the Lake Asphaltites. This position does not quite agree with the site of Petra, which is twice that distance from the Dead Sea, and about eighty-three Roman miles from Ailah or Akaba.

For the discovery and description of these interesting ruins geography is indebted to Burckhardt, who travelled through the mountains of Petræa in 1812. They were afterward visited (in 1818) by Captains Irby and Mangles, in company with Mr. Bankes and Mr. Legh; and more recently by two distinguished French travellers, MM. Leon de Laborde and Linant, whose talents have done for the tombs and temples of Petra what the splendid illustrations of Wood and Dawkins did for those of Palmyra.* The first sentiment that struck the mind of all these visiters was that of astonishment at the

* Voyage de l'Arabie Petrée, now in course of publication. Irby and Mangles's Travels, p. 407–437. Adventures of Giovanni Finati, vol. ii. chap. v. Macmichael's Journey to Constantinople, p. 228, &c: Seetzen passed through Idumæa in 1806, "where he expected to make several discoveries, but the fates decided otherwise." Zach's Corresp. p. 47. Burckh. Trav. in Syria, p. 422.

Entrance to Petra

utter desolation which now reigns over those once celebrated regions, described by an inspired pen as "the fatness of the earth." It is scarcely possible to imagine how a wilderness so dreary and desolate could ever have been adorned with walled cities, or inhabited for ages by a powerful and opulent people. The aspect of the surrounding country is singularly wild and fantastic. On one side stretches an immense desert of shifting sands, whose surface is covered with black flints, and broken by hillocks into innumerable undulations; on the other are rugged and insulated precipices, among which rises Mount Hor with its dark summits, and near it lies the ancient Petra, in a plain or hollow of unequal surface (Wady Mousa), enclosed on all sides with a vast amphitheatre of rocks.

The entrance to this celebrated metropolis is from the east, through a deep ravine called El Syk; and it is not easy to conceive any thing more awful or sublime than such an approach. The width in general is not more than sufficient for the passage of two horsemen abreast; through the bottom winds the stream that watered the city. As this rivulet must have been of great importance to the inhabitants, they seem to have bestowed much pains in protecting and regulating its course. The channel appears to have been covered by a stone pavement, vestiges of which yet remain; and, in several places, walls were constructed to give the current a proper direction, and prevent it from running to waste. Several grooves or beds branched off as the river descended, in order to convey a supply to the gardens, and higher parts of the city. On either hand of the ravine rises a wall of perpendicular rocks, varying from 400 to 700 feet in height, which often overhang to such a degree that, without their absolutely meeting, the sky is intercepted; scarcely leaving more light than in a cavern, for a hundred yards together. The sides of this romantic chasm, from which several small streamlets issue, are clothed with the tamarisk, the wild-fig, the oleander, and the caper-plant, which sometimes hang down from the cliffs and crevices in beautiful festoons, or grow about the path with a luxuriance that almost obstructs the passage. Near the entrance of the pass a bold arch is thrown across it at a great height. Whether this was the fragment of an aqueduct, or part of a road formerly connecting the opposite cliffs, the travellers had no opportunity of examining; but its appearance as they passed under it, was terrific, hanging over their heads between two rugged masses, apparently inaccessible. Without changing much its general direction, this natural defile presents so many windings in its course, that the eye sometimes cannot penetrate beyond a few paces forward, and is often puzzled to distinguish in what direction the passage will open. For nearly two miles its sides continue to increase in height as the path descends. The solitude is disturbed by the incessant screaming of eagles, hawks, owls, and ravens, soaring above in considerable numbers; apparently amazed at strangers invading their lonely habitation. At every step the scenery discovers new and more remarkable features; a stronger light begins to break through the sombre perspective; until at length the ruins of the city burst on the view of the astonished traveller in their full grandeur; shut in on every side by barren craggy precipices, from which numerous recesses and narrow valleys branch out in all directions; ending in a sort of *cul-de-sac*, without any outlet.

It was doubtless the impregnable nature of the place that rendered it so celebrated as a commercial depot; for while it admitted of easy access to beasts of burden, it might defy the attacks of robbers or enemies, however formidable. Though well chosen in point of security, the position of the town was subject to many inconveniences. The summer heats must have been excessive; as the bare elevated rocks, while they excluded the cooling breezes, would concentrate the sun's rays with double intensity. Safety and protection appear to have been the only objects that could induce a wealthy people to select so remarkable a site for their capital. The entire face of the cliffs, and sides of the mountains, are covered with an endless variety of excavated tombs, private dwellings, and public buildings; presenting altogether a spectacle to which nothing perhaps is analogous in any other part of the world. "It is impossible," says Mr. Banks, "to give the reader an idea of the singular effect of rocks tinted with the most extraordinary hues, whose summits present nature in her most savage and romantic form; while their bases are worked out in all the symmetry and regularity of art, with colonnades, and pediments, and ranges of corridors, adhering to the perpendicular surface." The inner and wider extremity of the circuitous defile by which the city is approached is sculptured and excavated in a singular manner; and these become more frequent on both sides, until at last it has the appearance of a continued street of tombs.

About half-way through there is a single spot, abrupt and precipitous, where the area of this natural chasm spreads a little, and sweeps into an irregular circle. This had been chosen for the site of the most elaborate, if not the most extensive, of all these architectural monuments. The natives gave it the name of Kazr Faraoun, the castle or palace of Pharaoh, though it resembled more the sepulchre than the residence of a prince. On its summit was placed a large vase, once furnished apparently with handles of metal, and supposed by the Arabs to be filled with coins; hence they denominated this mysterious urn the Treasury of Pharaoh. Its height and position have most probably baffled every approach of avarice or curiosity; from above it is rendered as inaccessible by the bold projection of the rough rocks, as it is from below by the smoothness of the polished surface. The front of the mausoleum itself rises in several stories to the height of sixty or seventy feet; ornamented with columns, rich friezes, pediments, and large figures of horses and men. The interior consists of a chamber sixteen paces square and about twenty-five feet high; the walls and roof are quite smooth, and without the smallest decoration. The surprising effect of the whole is heightened by the situation, and the strangeness of the approach. Half seen at first through the dim and narrow opening, columns, statues, and cornices gradually appear as if fresh from the chisel, without the tints or weather-stains of age, and executed in stone of a pale rose-colour. This splendid architectural elevation has been so contrived, that a statue, perhaps of Victory, with expanded wings, just fills the centre of the aperture in front, which, being closed below by the ledges of the rocks folding over each other, gives to the figure the appearance of being suspended in the air at a considerable height; the ruggedness of the cliffs beneath setting off the sculpture to the greatest advantage. No part of this stupendous temple is built, the whole being hewn from the solid rock; and its minutest embellishments, wherever the hand of man has not purposely effaced them, are so perfect, that it may be doubted whether any work of the ancients, except perhaps some on the banks of the Nile, has survived with so little injury from the lapse of time. There is scarcely a building in England of forty years' standing so fresh and well preserved in its architectural decorations as the Kazr Faraoun, which Burckhardt represents as one of the most elegant remains of antiquity he had found in Syria.

The ruins of the city itself open on the view with singular effect, after winding two or three miles through the dark ravine. Tombs present themselves not only in every avenue within it, and on every precipice that surrounds it, but even intermixed almost promiscuously with its public and domestic edifices; so that Petra has been truly denominated one vast necropolis. It contains above two hundred and fifty sepulchres, which are occasionally excavated in tiers, one above the other; and in places where the side of the cliff is so perpendicular that it seems impossible to approach the uppermost, no access whatever being visible. There are besides numerous mausoleums of colossal dimensions, and in a state of wonderful preservation. Near the west end of the wady are the remains of a stately edifice, the Kazr Benit Faraoun, or palace of Pharaoh's daughter, of which only a part of the wall is left standing. Towards the middle of the valley, on the south side, are two large truncated pyramids, and a theatre, with complete rows of benches capable of containing above 3000 spectators, all cut out of the solid rock. The ground is covered with heaps of hewn stones, foundations of buildings, fragments of pillars, and vestiges of paved streets,—the sad memorials of departed greatness. On the left bank of the river is a rising ground, extending westward for about three-quarters of a mile, entirely strewn with similar relics. On the right bank, where the ground is more elevated, ruins of the same description are to be seen. In the eastern cliff there are upwards of fifty separate sepulchres close to each other. There are also the remains of a palace and several temples; grottoes in vast numbers, not sepulchral; niches, sometimes excavated to the height of thirty feet, with altars for votive offerings, or with pyramids, columns, and obelisks; horizontal grooves for the conveyance of water, cut along the face of the rock, and even across the architectural parts of some of the excavations; dwellings scooped out, of large dimensions, in one of which is a single chamber sixty feet in length and of a proportional breadth; many other habitations of inferior note, particularly numerous in one recess of the city, the steep sides of which contain a sort of excavated suburb, accessible only by flights of steps chiselled out of the rock. In short, the outer surface of the strong girdle that encircles the place is hollowed out into innumerable artificial chambers of different dimensions, whose entrances are variously, richly, and often fantastically decorated with every order of architecture; showing how the pride and labour of art has tried to vie with the sublimity of nature. The effect of the whole is heightened by the appearance of Mount Hor towering above this city of sepulchres, and perforated almost to the top with natural caverns and excavations for the dead.

The immense number of these stupendous ruins corroborates the accounts given, both by sacred and profane writers, of the kings of Petra, their courtly grandeur, and their ancient and long-continued royalty. Great must have been the opulence of a capital that could dedicate such monuments to the memory of its rulers. Its magnificence can only be explained by a reference to the immense trade of which it was the common centre from the dawn of civilization. The structure of many of these edifices denotes pretty nearly the age to which they belong. Their relics exhibit a mixture of Grecian and Roman architecture, although the ground is strewn with others of a more ancient date. Among the views given by the French travellers is one of a tomb on which is engraven a Latin inscription, with the name of a magistrate, Quintus Pretextus Florentinus, who died in that city, being governor of Palestina Tertia about the time of Adrian or Antonine (A. D. 126–160). These writers have illustrated and described another splendid monument,—a temple situated westward, on the bank of the river,—the only edifice in that spot which has resisted the ravages of time. The outlines of its beautiful architecture are tolerably perfect; and the cornice which surmounts the wall is in a pure and elegant style.

These magnificent remains can now be regarded

only as the grace of Idumæa, in which its former wealth and splendour lie interred. The state of desolation into which it has long fallen is not only the work of time but the fulfilment of prophecy, which foretold that wisdom and understanding should perish out of Mount Seir; that Edom should be a wilderness; its cities a perpetual waste, the abode of every unclean beast. "Thorns shall come up in her palaces, nettles and brambles in the fortresses thereof; the cormorant and the bittern shall possess it, and it shall be an habitation of dragons, and a court for owls. The wild beasts of the desert shall also meet with the wild beasts of the island, and the satyr shall cry to his fellow; there shall the vultures also be gathered, every one with her mate; there shall the screech-owl make her nest and lay, and find for herself a place of rest." (Isaiah xxxiv. 5, 10, 17.) Nowhere is there a more striking and visible demonstration of the truth of these divine predictions than among the fallen columns and deserted palaces of Petra. The dwellers in the clefts of the rocks are brought low; the princes of Edom are as nothing; its eighteen cities are swept away, or reduced to empty chambers and naked walls; and the territory of the descendants of Esau affords as miraculous a proof of the inspiration of Scripture history as the fate of the children of Israel.*

* Keith's Evidence of Prophecy.

CHAPTER V.

Character, Manners, and Customs of the Ancient Arabs.

Two Classes of Arabs—The Bedouins, or Pastoral Tribes—Their Mode of Life—Their Love of Freedom—The Agricultural and Mercantile Classes—Commerce of the Ancient Arabs—Their early Intercourse with India—Wealth and Luxury of the Sabæans—Exaggerated Accounts of the Greeks and Romans—Neither Gold nor Silver Mines in Arabia—Principal Articles of Trade—Frankincense—Myrrh—Coffee—Vines—Sugar—Chief Marts on the Coast—Caravans—Propensity of the Arabs for Robbery and War—Their Quarrels and Revenge—Their sacred Months—Their Hospitality—Fire-signals—Liberality of Hatim Tai—Fondness of the Bedouins for Eloquence and Poetry—The Moallakat, or Seven Poems of the Kaaba—Origin and Copiousness of the Arabic Language—Learning and Morals of the Ancient Arabs—Their Division of Time—Their Superstitions—Charms—Sortilege—Divination by Arrows—Worship of the Stars and Planets—Popular Idols and Images—Planting of Christianity in Arabia—Labours of Origen—Bishops' Sees—Schisms and Heresies in the Arabian and Eastern Churches.

The distinction of two great classes is as strongly marked in the character and habits of the Arabs as in their genealogical descent. The natives of the desert, who follow a pastoral and predatory life, consider themselves as a separate race from the inhabitants of cities and towns, who live by tillage and commerce. The former have a variety of names by which they designate themselves, all expressive of their peculiar mode of life. They are called Ahl el Hajr, or the People of the Rock; Ahl el Wabar, the Dwellers in Tents; Bedawiyun (Bedouins), the Inhabitants of the Desert, &c. All the other classes who are fixed in local habitations, or engaged in the pursuits of industry, they stigmatize as Ahl el Madar the Dwellers in Houses made of clay. Through all antiquity this characteristic distinction has remained inviolate; and it continues in force at the present day, as strongly marked as it was three or four thousand years ago.

The Scenite or Nomadic tribes held in contempt the peaceful and mechanical arts; and had any of their number abandoned their erratic habits for the occupations of agriculture, they would have been considered as degraded, and fallen from the primitive nobility of their birth. Their grand employment was the tending of their flocks, which constituted their principal wealth, and supplied all their domestic necessities. They held little intercourse and had few connexions with the world around them; but their habits of sobriety raised them above the artificial wants of more refined and civilized nations. It was their constant boast, that little was required to maintain a man who lived after the Bedouin fashion. Their chief nourishment was dates and milk. The camel, the most common and the most valuable of their possessions, was of itself a storehouse of useful commodities. The flesh of the young was tender, though reckoned conducive to a hot and vindictive temperament; the dung was consumed as fuel; the long hair, which fell off annually, was manufactured into curtains for their tents, and various articles of dress and furniture.

While food and raiment were thus supplied by the spontaneous gift of nature, they envied not the tenants of the more fertile and industrious provinces. Their love of liberty was stronger than the desire of wealth; and the passion for foreign luxuries, which has proved so fatal to other countries, has not yet changed the patriarchal manners of the roving shepherds of Arabia. As all travellers have remarked, the modern Bedouin differs but little from his ancestors, who, in the age of Moses and Mohammed, dwelt under similar tents, and conducted their flocks to the same springs and the same pastures. It is in the lonely wilderness and the rugged mountains that his attachments centre; because it is there he can live without ceremony and without control. The very wildness of this inhospitable scenery constitutes in his eyes its principal charm; and were these features destroyed, the spell would be broken that associates them in his mind with the romantic freedom of his condition. The tent he regards as the nursery of every noble quality, and the desert the only residence worthy of a man who aspires to be the unfettered master of his actions. He cannot imagine how existence can be endured, much less enjoyed, except in a dwelling of goats hair, which he can pitch and transplant at pleasure. These are privileges which he would not exchange for rubies. His steril sands are dearer to him than the spicy regions of the south; and he would consider the security of cities but a poor compensation for the loss of his independence. It was an ancient proverb, of which the Arabs made their boast, that God had bestowed on their nation four precious gifts. He had given them turbans instead of diadems, tents in place of walls and bulwarks, swords instead of intrenchments, and poems instead of written laws.

This state of uncontrolled existence has in all ages been the object of their wishes and their pride; and it never has been renounced without profound regret. Abulfeda has preserved a very lively trait of this feeling in the complaint of Maisuna, an Arab lady married to one of the caliphs of Damascus. The pomp and splendour of an imperial court could neither reconcile her to the luxuries of the harem, nor make her forget the homely charms of her native wilderness. Her solitary hours were consumed in melancholy musings; and her greatest delight was in singing the simple pleasures she had enjoyed in the desert. The modern Bedouins decline the shel ter of houses when business calls them to visit crowded cities. They are seen passing the night in the gardens or public squares of Cairo, Mecca, and Aleppo, in preference to the apartments that are offered for their accommodation.* These local attachments seem strongest in the inhabitants of mountainous countries. The Scottish Highlander, wherever he roams, thinks with pleasing regret on his dark hills. The exiled Swiss pines for his bleak Alps, and the wild melody of his native songs. The Laplander has fixed the site of the terrestrial paradise amid his own dreary wastes. The boatmen on the Nile lighten the cares of bondage or banishment by singing "Nubia is the land of roses!" The Druse on the rugged summits of Lebanon looks down with indifference on the blooming valleys of the Jordan, that spread their enchanting beauties at his feet. The same feeling glows with more than ordinary warmth in the bosom of an Arab; and in preferring the rude simplicity of his paternal solitudes to the comforts and luxuries of more refined society, he yields only to a common but a kindly instinct of human nature.

In some parts of the northern deserts there were migratory tribes not entirely addicted to the pastoral life. Whether from the advantages of a less steril soil, or the vicinity of Palestine and Syria, or the example of the emigrants from Yemen, they were distinguished from their central brethren by their residence in towns, and their application to the arts.

* Abulfed. Annal. Moslem, a Reiske, vol. i. p. 116. Reynier, de l'Economie Pub. et Rur. des Arabes, p. 25. Prof. Carlyle has translated this fragment in his Specimens of Arabian Poetry, p. 31. The same feeling inspired other royal bosoms than Maisuna's.

———————" Nebassar's queen,
Fatigued with Babylonia's level plains,
Sigh'd for her Median home, where Nature's hand
Had scoop'd the vale, and clothed the mountain's side
With many a verdant wood."—Robert's *Judah Restored.*

Some of them were entirely occupied in agriculture, while others added to the toils and pursuits of a sedentary life a taste for pillage; and divided their number so, that while one part attended to domestic labours, the others were engaged in war and plunder. Such was their condition until overrun by the Romans, whose dominion effaced the last vestiges of their industry, laid their cities in ruins, and reduced their territories to a state of desolation from which they have never recovered. Though these half-civilized tribes shared with their wild clansmen in the interior the same warlike propensities, they had not the same facilities of withdrawing from danger; consequently their liberty was more precarious. When assailed by the neighbouring nations, they could purchase security only by submission or tribute,—which was always exacted, whether the sovereigns of the East or the West were their masters. Among the nations that paid this ransom to Jehoshaphat, king of Judah, are mentioned the Arabs—evidently the pastoral tribes of which we are speaking—who paid annually 7700 sheep, and as many goats (2 Chron. xvii. 11). It was the danger or the necessity of yielding to the mercenary power of tyrants, that confirmed the nomadic Arabs in their dislike of settled habitations. The Chevalier D'Arvieux observes, that their attachment to the wandering life proceeded from their notion that it was more congenial to liberty; since the shepherd who ranges the desert with his herds will be far less liable or likely to submit to oppression than the proprietor of houses and lands. A passage in Diodorus shows how ancient and deeply rooted was this mode of thinking among the Bedouins. "The Nabathæans," as he calls them, "were prohibited by their laws from sowing, planting, drinking wine, and building houses. Every violation of this statute," he adds, "was punishable by death." The same was the case with the Rechabites, an Arab tribe mentioned by Jeremiah (chap. xxxv. 1–11), who adhered to this law of their country long after they had become resident in Judæa.* The reason assigned for this interdict was, the belief that the possessors of fields and vineyards would be more easily brought under subjection to a foreign yoke.

Circumstances so dissimilar between the wandering and the stationary tribes could not fail to induce habits essentially different. Among the latter the subdivision of the soil was adopted, and consequently the right of individual property respected. That they had made very considerable progress in agriculture is certain; but of the theory and practice of their husbandry we must be content to remain in ignorance. Their principal occupation—to which they owed their wealth and their fame—was commerce. That the Arabs were the first navigators of their own seas, and the first carriers of oriental produce, is evident from all history; and that they had been so from the remotest ages we may safely infer from analogy, from necessity, and from local situation. Sabæa, Hadramaut, and Oman were the residence of merchants from the very dawn of civilization. They had frequented the ports of the Red Sea, crossed the Persian Gulf, and, with the aid of the monsoons, visited the coasts of India long before these regions were known to the nations of Europe. Moses speaks of cinnamon, cassia, myrrh, and other aromatics appropriated to religious uses (Exodus xxx. 22–25); and he mentions them in such quantities, as plainly shows they were neither of rare nor difficult attainment; and that even in his time the communication was opened between India and Arabia. It was to this source that ancient Egypt owed its wealth and its splendour. Thebes and Memphis traded with the Arabs, and were celebrated as mercantile cities more than a thousand years before the foundation of Cairo or Alexandria. In

* Townson's Hist. of the Rechabites. Wolff's Journal.

the hands of Solomon the traffic of the Red Sea produced a revenue equivalent to three millions and a half (1 Kings x. 14); and at the period when the Romans invaded the East, this lucrative monopoly was exercised by the Sabæans, whose marts they found richly stored with all the precious commodities of India.

All writers, sacred and profane, speak of the valuable and extensive trade of Saba or Yemen. The portrait which Ezekiel (chap. xxvii.) draws of Tyre is not only curious as an illustration of ancient commerce, but may be regarded as conveying a faithful description of the mercantile activity which the universal intercourse of nations must have created in the seas and harbours of Arabia. "Tarshish was thy merchant by reason of the multitude of all kind of riches; with silver, iron, tin, and lead they traded in thy fairs. The men of Dedan were thy merchants in precious cloths for chariots. Syria was thy merchant by reason of the multitude of the wares of thy making: they occupied in thy fairs with emeralds, purple, and broidered work, and fine linen, and coral, and agate. Dan and Javan, going to and fro, occupied in thy fairs: bright iron, cassia, and calamus were in thy market." Job alludes to the pearls and rubies, the precious onyx, the sapphire, the coral, and the topaz; which shows that even at his early age the northern tribes were not ignorant of the luxuries of their more wealthy neighbours. Diodorus considered the Happy Arabia so immensely opulent, that all the treasures of the world seemed to centre there as in one universal mart. Agatharcides, the first historian worthy of credit who describes the commerce of Yemen, and its different productions, either native or imported, has given a singular picture of oriental trade as it stood in the reign of Ptolemy Philomater, nearly two hundred years before the Christian era. At that time Arabia was the medium of communi cation between India and Egypt, and it was in her ports that the Greeks were wont to purchase their cargoes, before they had ventured to make the distant voyage themselves. Saba, he observes, abounded with every production that could make life happy in the extreme. The land yielded not merely the usual commodities; balm and cassia, incense, myrrh, and cinnamon were of common growth. The trees wept odorous gums, and the gales were so perfumed with excessive fragrance that the natives were obliged to renew their cloyed sense of pleasure by burning pitch and goats' hair under their noses. They cooked their victuals with scented woods; living in the careless and delightful enjoyment of those blessings which conferred on their country the appellation of Happy. In their expensive habits they rivalled the magnificence of princes. Their houses were decorated with pillars glistening with gold and silver. Their doors were of ivory, crowned with vases and studded with jewels. The interior of their habitations corresponded with their outward appearance; in articles of plate and sculpture; in furniture, beds, tripods, and various household embellishments, they far surpassed any thing that Europeans ever beheld.

Other writers speak in similar terms of the luxury and riches of the Sabæans. Arrian, in the Periplus, mentions their embroidered mantles, their murrhine vases, their vessels of gold and silver elegantly wrought, their girdles, armlets, and other female ornaments. Strabo describes their bracelets and necklaces, made of gold and pellucid gems arranged alternately; as well as their cups and other domestic utensils, all composed of the same precious metals, which we are assured were so abundant that gold was but thrice the value of brass, and only twice that of iron; while silver was reckoned ten times more valuable than gold;—their mountains producing the latter commodity in vast quantities nearly in its pure state, and in lumps from the size of an olive to that of a nut.* Diodorus has a statement to the same effect. Through the whole country of the Debæ (in Hejaz), he tells us, passed a river so abounding with small gobbets of gold, that the mud at its mouth seemed to consist almost entirely of that metal, which was of so bright and glorious a colour that it added an exceeding lustre and beauty to the most valuable gems that were set in it.

Such is the brilliant picture which the enraptured imagination of foreigners drew of the Happy Arabia. It were almost unkind to disenchant the reader of these golden visions; but truth compels us to avow, that the discoveries of modern travellers have drawn aside the veil of romance from this fairy-land. They do not find it that paradise of ambrosial felicity and inexpressible delights which antiquity represented it. Its real or ideal treasures have vanished; and no Alexander in our times would dream of making its balmy vales the seat of his mighty empire. Old Ocean, to use Milton's simile, is no longer cheered with the grateful smell of the spicy odours that are blown by the north winds from the Sabæan coast.† Many valuable articles which, under the Ptolemys and the Cæsars, were considered as the productions of Arabia, were afterward found to be imported from India, Caramania, and Serica. Even the boasted incense, which Pliny treats at length in the twelfth book of his Natural History, is not wholly the indigenous gift of that country. The kind called

* This cheapness of gold may appear incredible; but the Mexicans and Peruvians seem to have held it in as little esteem as the ancient Arabs: they exchanged it readily with the Spaniards for knives, hatchets, glass beads, and other trifles.

† ——————"Many a league
Cheer'd with the grateful smell, old Ocean smiles."
Parad. Lost, book iv.

The writer of an old history of the Turkish Empire says, 'The air of Egypt sometimes in summer is like any sweet perfume, and almost suffocates the spirits; caused by the wind that brings the odours of the Arabian spices."

olibanum by the natives, the quality of which is very inferior, is the only species cultivated by them, and is produced in the province of Sahar, and the mountains of Dhafar and Merbat. They procure several kinds from Abyssinia, Sumatra, Java, and Siam, which they export in great quantities to Turkey.

According to Niebuhr's account, neither gold nor silver mines are worked or even known in Arabia; though a small quantity of the latter metal is extracted from the lead obtained in the province of Oman. The onyx, a kind of emerald employed in adorning the walls of houses; amber, blue alabaster, selenite, and various spars, are found in Yemen; though the Danish traveller mentions, that most of the precious stones ascribed to that country are brought from India. Still, though shorn of its honours, it is difficult to efface from the mind the glowing descriptions of antiquity. These must have rested on a solid basis of truth, clouded as they may have been with fable and hyperbole. The positive and unanimous testimony of so many writers cannot allow us to doubt their accounts, though we may suspect exaggeration. That the mountains of Yemen once yielded gold, sometimes found in the body of the rocks, or in loose globules on the surface, or in the sands of the rivers, is an historical fact to which there appears no reason whatever for refusing our assent. The evidence that Solomon obtained gold from Arabia is express. Without attaching much importance to the thuriferous encomiums of Virgil and Horace, we may presume that the legions of Greece and Rome would not have braved so many dangers, nor courted an alliance with the Sabæans, had the report of their wealth been an idle fiction.*

* "India mittit ebur, molles sua thura Sabæi."—*Virg.* Georg. i. l. 57. Horace indeed thinks his friend Mæcenas should prefer a lock of Licinia's hair to the "plenas Arabum domos."—*Carmin.* lib. ii. ode 12. Philostorgius, Hist. Eccles. lib. iii. c. 4.

With regard to the exports of the ancient Arabs, these consisted chiefly of the productions of their own soil, whether natural or reared by cultivation. The articles of the transit trade were more various, and have been enumerated by Arrian, Agatharcides, and other authors. In these invoices are found gold, silver, iron, lead, tin, brass, of which were made drinking-vessels, cooking-utensils, and ornaments of all sorts; ivory, tortoise-shell, flint-glass, carved images, javelins, hatchets (πελυκια,), adzes (σκεπαρνα) knives, awls or bodkins (όπητια), cloths of various kinds, military cloaks (άβολλαι), tartans (διкροσσια) died mantles for the Berberine markets in the African ports of the Red Sea; fine muslins, silks, linens, quilts or coverlids (λωδικες), manufactured at Moosa; coarse cottons (μολοχινα), girdles, long sashes, died rugs made at Arsinoë; Chinese furs, and female dresses of every description. Of gums, spices, and gems, the varieties are numerous; bdellium, cinnamon, ginger, cassia, of which ten sorts are specified in the Periplus; honey, spikenard, sugar, pepper, stibium for tinging the eyelids black, and storax, one of the most agreeable of the odoriferous resins. The apes and peacocks mentioned in Scripture, as well as many of the spices and precious stones, were undoubtedly the produce of India, and brought by the Arabs into Palestine.

Among the principal articles of native growth must be ranked the incense so famous in all antiquity. Its use in religious oblations, among the Jews and other oriental nations, ascends to an era infinitely remote. The various offerings of the Israelites, even in the wilderness, were perfumed with it; from their altars night and morn, which were never quenched, its odorous exhalations ascended perpetually for a sweet-smelling sacrifice. "To what purpose," says Jeremiah, remonstrating with his idolatrous countrymen, "cometh there to me incense from Sheba, and the sweet cane from a far country? your burnt-offerings are not acceptable, nor your sacrifices sweet unto me." (Chap. vi. 20.) Theophrastus speaks of the vast quantities that were collected and brought from every part to the temples of the Sun. The deity worshipped at Saba exacted a tithe of all the incense, which was brought in immense quantities on the backs of camels. Virgil speaks of the hundred altars that smoked with Sabæan incense in honour of the Paphian Venus. Homer, as has been often remarked, makes no allusion to this substance, though it is repeatedly mentioned by Herodotus.

All that we find in ancient authors, regarding the production and collection of this celebrated commodity, is wrapped in fable and mystery. Naturalists are not even certain as to the kind of shrub that produced it; but it is supposed to be a species of the amyris or balm-tree, so rich in affording gum-resins. Theophrastus, in his History of Plants, says it grew wild in Arabia on the slopes of the mountains, and that it was also cultivated on the plains surrounding their base. The native writers assert, that it was only found in the eastern part of the Happy regions; and that the districts where it was most common were those of Merbat, Sahar, and Mahrah, though the latter had neither fruit nor palm-trees. Hadramaut and an extensive tract called Sachalites are mentioned by the Greek geographers as the native country of incense; which they represent as of difficult access, extremely unhealthy from the thickness of the air, and grievously infested by serpents, whose bite was incurable. These reptiles are described as very fierce, of a small spotted kind, and so numerous around each tree, that the peasants were obliged to burn stryax in order to expel them. When gathering the incense, the people spread mats or stone slabs under the branches during the seasons when they scarified or made incisions in the bark, and this happened twice annually,—in winter and about midsummer. The former collection was reckoned of inferior quality.*

Theophrastus and Pliny after him relate, that all the thuriferous trees were the property of 300 families, who were reputed holy; each having their separate allotment, and reserving a portion of the harvest for their gods. But as neither of these writers travelled in the country, their reports do not merit implicit credit. The author of the Erythræan Periplus, who had navigated this sea and visited the coasts of Arabia, takes no notice of those sacred families, but he has recorded the stories told him by others. The incense, he informs us, grew in places so unhealthy, that nobody but slaves and malefactors were employed in gathering it; that it belonged entirely to the government, and was so immediately under the protection of the gods, that certain destruction was sure to overtake all who should dare to procure it by contraband means. The terror of serpents and offended divinities was probably an invention of the Arabs to protect their commerce from foreign intrusion, and to frighten strangers away from the places where the balm-trees grew, that they might preserve the monopoly of incense to themselves.

Myrrh was also one of the productions of Yemen, but we are assured by Strabo, Arrian, and Diodorus that it was obtained in greater quantities and of a superior quality in the neighbouring countries of Africa. The kind of tree that yielded it has not been exactly described, but there is little doubt that it was a species of the amyris. The ancients inform us, that its branches required frequent pruning, and that it was nowhere found in its wild state in Arabia,

* Pliny's account of the incense is throughout extremely curious. Thousands of families were employed at Alexandria in preparing it. The workmen, he says, wore masks, and were stripped naked, to prevent them from abstracting the smallest particle.—Lib. xii. c. 14.

out demanded constant cultivation; facts which would serve to indicate that it was not indigenous to the soil, but imported from the West. Like the incense, it exuded from the bark; the incisions were made at the same periods of the year, and the summer product was reckoned the best.

Besides these two substances, which formed the principal commercial articles of Arabian exportation, there were others of a secondary importance; such as benzoin, an oil which, from the most remote antiquity, has been preferred to all others in the preparation of perfumes; the aloes of our pharmacopeias, the best kind of which grows in the isle of Socotra; balm, which the Arabs furnished long before the Jews had introduced into Palestine the culture of the tree from which it is obtained. Coffee, so famous over all the world, and from which the natives of Yemen now derive a vast revenue, does not appear to have always existed or been cultivated in that country. The ancients at least knew nothing of it; nor do we find it mentioned in the earlier Mohammedan writers. It does not appear to grow wild in Arabia; but it is found in that state in several districts of Africa, where the inhabitants use it as food in their military expeditions. It is not improbable that the Abyssinians, who made use of coffee when they invaded Yemen, might introduce that tree, and cultivate it so long as they remained masters of the country. The monopoly of this article would at first remain in their hands; but when Mohammed discommended, if he did not absolutely proscribe, the use of fermented liquors, such a privation might suggest the idea of substituting a decoction of this nutritive berry.

The soil of the greater part of Arabia was unfavourable to the vine. The juice of the grape was freely allowed; and the scarcity of that beverage may be inferred from a passage in Arrian, who recommends strangers to bring, among other presents, wine for the use of the native kings and chiefs. Abulfeda says, that the inhabitants in the vicinity of the Dead Sea coated the trunks of their vines and palm-trees with bitumen, thinking its tendency beneficial; but this was a local practice rather than a general system. The ancients say that the sugar-cane was also cultivated there, but that its fruit was of inferior quality to that produced in India. As Arrian speaks only of Indian sugar, we may conclude that Arabia furnished, or at least exported, but little. The notions of these writers as to this article were extremely imperfect; they supposed it to be crystals naturally formed in a species of reed; and the small quantity that was brought to Europe was employed in medicinal prescriptions. It is to the Arabs, however, that we owe the introduction and cultivation of this valuable plant.

Of the different countries that traded with Arabia,—of the rates of exchange, or the relative amount of exports and imports, we are left without any precise information. That the balance of trade must have been in favour of the Arabs, we may infer from the rich dresses and gorgeous luxuries which attracted the wonder and cupidity of all strangers. The vast intercourse of merchants, while it opened up a ready sale for the different productions of their own country, enabled them to realize immense profit in bartering the various articles for which they were the principal agents. From their geographical position, they became the natural centre of all the traffic between India, Africa, and Europe. On the western shore of the Red Sea the chief marts were Arsinoë, Myos Hormus, which D'Anville places in 27° north latitude, Berenice, Ptolemaïs Theron and Adulis. It was in the two latter ports that the hunters of Ptolemy procured elephants for his army. On the Arabian coast, the harbours most frequented were Ælana or Ezion-gaber; Leuké Komé, where the Romans in the time of Augustus had a garrison

Moosa, more than a thousand miles down the gulf; and Ocelis on the Strait of Diræ (Bab el Mandeb). The latter market in the time of Pliny was the most celebrated for the merchandise of India, while Moosa was the great *entrepôt* for the native productions of Arabia. Aden was the ancient centre of traffic between India and the Red Sea. The ships from the East, being too large to pass the strait, landed their cargoes here. It was destroyed by the Emperor Claudius, with the view of suppressing every power that might interfere with the Roman commerce, and of increasing his revenues from the duties on the trade of the gulf, which were collected by Plocamus, with the garrison stationed at Leuké Komé. In the time of Constantine, Aden had recovered its commercial celebrity; and, being subject to the empire, it got the name of Romanum Emporium. Kané, Sahar (the Sachalites of the Greeks), and Moscha (Muscat) were noted for various native exports, especially incense and aloes. Gherra was the most celebrated mart on the Persian Gulf. Pliny says the city was five miles in circumference, with walls and towers built of fossil salt. Agatharcides compares their riches with those of the Sabæans; and adds, that they brought much wealth into Syria.

The commerce of the Arabs was not confined to the maritime towns. The difficult navigation of their seas made it necessary to employ transports by land for a considerable portion of their merchandise. Trade was in this manner compelled to take a variety of directions across the peninsula, which from time immemorial was traversed by numerous caravans, establishing a regular communication from sea to sea, and connecting distant countries with each other. The Gherreans crossed the desert to Petra, Bosra, and Damascus. They were among the earliest conductors of caravans on record. We learn from Strabo, that they furnished a variety of articles in spices and aromatics to the Tyrians and Pheni cians, who commanded the trade of Egypt and the Mediterranean. He mentions another caravan that departed from Ælana, performing a journey of seventy days to the Land of Incense. The Koreish, as we have observed, were early distinguished for their mercantile enterprise, to which their central position was highly advantageous. Their camels travelled to Sanaa, and returned with the precious cargoes of Moosa and Aden, which they exchanged for grain and provisions in the Syrian markets. Even at that remote period foreign merchants appear to have brought their native productions to Mecca during the annual festivals. Agatharcides and Diodorus seem to allude to this religious traffic when they speak of a fair in Arabia, which was frequented by different nations for the purpose of trade.

Political circumstances necessarily affected these inland conveyances, and gave a greater or less degree of activity to commerce, according to the prosperity or decline of the countries with which it was carried on. When the Phenicians ceased to be a maritime power, their place was supplied by Egypt, which, under the first Ptolemys, had risen into new life and vigour. The caravans that formerly arrived at Tyre then took the route to Alexandria, long a celebrated *entrepôt* of oriental trade. When the Romans conquered that province, the lucrative traffic fell into their hands; and their mercantile industry in the Eastern seas contrasts singularly with the desolating spirit which everywhere else marked the progress of their arms. Their taste for foreign rarities became a passion; and drew from Pliny the complaint, that the empire was exhausted by a drain of 400,000*l.* a year, required for the purchase of articles equally expensive and superfluous. Horace rallies his friend Iccius on being so smitten with the common rage for Arabian luxuries as to abandon the study of philosophy for the profession of arms, in the hope of conquering that envied but indomitable country. When the Roman power yielded in its turn to the victorious Saracens, the trade of the East reverted to the Arabs; and with them it might have still remained had no Gama arisen to double the Cape of Storms, and effect a total revolution in the whole commercial system of the world. The recent intercourse with Europe has introduced still greater changes; but when the Portuguese first visited the Arabian seas, they found the country and the commerce nearly in the same state as had been described by the Greeks and Romans 1500 years before.

The national character of the inhabitants was little affected by their external connexions. Their love of independence was attended with certain baneful effects on society. It engendered pride, and nursed into action those tendencies to war, cruelty, and rapine which they inherited from nature. Cherishing an unsocial disdain towards all mankind, their hostilities recognised no distinctions; their only rule was their own advantage; and whenever this was concerned, they attacked friends and foes without scruple or provocation. As they made no difference between war and pillage, robbery by armed force was confounded with the rights of conquest. The plundering of a solitary traveller was in their eyes as much a military exploit as the sacking of a town or the reduction of a province. The poverty of their own land was with them an honourable excuse for relieving their wants at the expense of their wealthier neighbours. They alleged, that in the division of the earth the rich and fertile parts were assigned to the other branches of the human family; and that the descendants of the outlaw, "whose hand was to be against every man," might recover by force the hereditary portion of which they had been unjustly deprived. Though the posterity of Isaac and Abraham, strictly speaking, were the only people from whom they should have sought indemnity, yet their rapacious spirit easily extended this license to other nations who could not be accused of usurping their patrimony.

This fierce temperament, which waged war against all the world, was in no degree mitigated in their own domestic broils. Though the strictest honour and probity reigned amid their tents when their passions were unruffled, their wrath on provocation burned with double fury; and in their sanguinary feuds, the voice of law and humanity was alike disregarded. In the times of which we now write more than two thousand battles have been recorded. These combats generally took place between different families or clans; they often rose from trivial causes, but seldom ended without deeds of revolting atrocity. The war of the two horses, Dahes and Ghabra, between the tribes of Aus (or Abs) and Dobian, occasioned by a contested race, lasted forty years; during which period all industry was at a stand, and thousands were either slain in pitched battles or privately assassinated. The war of Basus sprang from the shooting of a camel which had drunk at a forbidden spring. It raged many years between the tribes of Becr and Taglab, until nearly all the principal men on both sides were cut off. A contemptuous word, an indecent action, an insult offered to their women or their beard, could only be expiated by the blood of the offender. The war of Nefravat, which set nearly the whole kingdom of Gassan in a flame, took its origin from the upsetting of an old woman, who had brought a basket of tributary butter to one of the chiefs, the quality of which happened to displease his fastidious palate. The most wanton cruelty was sometimes inflicted on the captives taken in these contests. Shaerhabil made them walk barefooted over heated stones, insulting their miseries by promising to buy them new soles.*

* Rasmussen, Additam. ad Hist. Præcip. Reg. Arab. Notæ. Pococke Specimen, p. 48.

The Arabs were a most vindictive race; their quarrels were hereditary. A wilful offence was never forgiven; and such was the patient inveteracy of their wrath, that they would wait months and years the opportunity of revenge. Murder, in these cases, was atoned for by fine or confiscation; but the kinsman of the deceased was at liberty to accept the penalty, or exercise with his own hand the law of retaliation. Nor was the debt of vengeance cancelled even by the head of the murderer; for their refined malice could substitute the innocent for the guilty, and transfer the punishment to the most considerable of the tribe by whom they had been injured. The slightest provocation, an empty boast, or the recital of an obnoxious song was enough to rekindle the embers of discord, and awaken passions that had slumbered for half a century.

The only respite to these sanguinary feuds was an interval of four months each year—the first, seventh, eleventh, and twelfth, which were always observed as sacred. During that time they held it unlawful to wage war foreign or domestic. Their swords were religiously sheathed; the heads taken off their spears; the injured and the injurer lived in perfect security, so that if a man met the slayer of his father or his brother he durst offer him no violence. This singular institution was observed by all the Arab tribes except two—Tai and Katham: and so scrupulously was it adhered to, that history has recorded only four (or six) instances of its having been transgressed. These were stigmatized by the name of the Impious Wars; in some of which the Koreish themselves were engaged. The design of this jubilee is alleged to have been for the security of merchants and pilgrims, that they might travel to and from the annual festival without danger or interruption.

This dark side of the Arab character had a beautiful contrast in certain noble and generous quaitiles

The moment the fierce marauder ceased to be in a state of war, he became quite another man. His tent was the asylum of the stranger, the home of kindness and hospitality. The traveller who sought his protection, or confided in his honour, he entertained without inquiry or the hope of remuneration He regarded him, not merely as a guest, but as a member of his family; he would defend his life at the risk of his own; and dismiss him, after the enjoyment of needful repose, with blessings, perhaps with gifts. His word once pledged was a sure and inviolable guarantee. The ancients extolled the Arabs for their benevolence. Agatharcides represents them as the most hospitable people on earth. Treatises and poems were composed expressly to eulogize their superiority in this respect. The highest compliment that could be paid a man was to pass an encomium upon his munificence, and the most acceptable to a woman was to celebrate her parsimony and her beauty. The contrary vices were as much an object of contempt and reproach; and a certain poet, satirizing the inhabitants of Waset, upbraids them, as the deepest stain on their honour, that none of their men had the heart to give, nor their women to deny.

Their friendly treatment of strangers was not confined to the camp or the tent. On every hill the "fires of hospitality" nightly blazed, to conduct the wayfaring traveller to a place of safety and repose. Amid the darkness of winter, the country for miles round was lighted up with these beacons; and the higher and larger they were, the more honourable was the generosity esteemed of him that provided them. It was a matter of glory and rivalry to surpass each other in the number and extent of these kindly tokens. "Thy fires," says a poet, "are kindled after sunset in every valley. The weary traveller spies these red signals afar through the obscure night."

The emblem of fire was used by the Arabs in a variety of indications (Nuvairi mentions fourteen) as well as that of hospitality; and the purpose of each was as easily distinguished as the state of the tide is made known to the distant mariner by the ball and pennon on shore. If a guest was unwelcome, he was dismissed; "the fire of expulsion" was kindled behind him, and he was told that God wished him never more to return. "The fire of safety" told his tribe that the freebooter had returned secure with his plunder. "The fire of war" admonished the more remote inhabitants of their impending danger; and it was not uncommon to see the flames of peace and hostility rising from the same hill. "The fire of the lion" was used to scare away that troublesome and terrible animal: "the fire of hunting," to dazzle the eye of the gazelle, that it might be the more easily caught: "the fire of the brand" was the mark imprinted on the necks of their camels,—a ring, a needle, or a cross,—by which each man knew his own, and might recognise them if stolen. "What is thy fire or brand?" was the question usually put to the person inquiring after cattle or offering them for sale to the merchant. "The fire of the oaths" was that over which every league or compact was solemnized. Into this they threw sulphur and salt, to inspire terror; and, recounting the various blessings derived from that element, they invoked the Deity to withdraw them from the breaker of his pledge; and if any had so violated his engagement, his treachery was proclaimed by "the fire of perfidy."

The virtue of hospitality often degenerated into foolish extravagance; and there were individuals who strove to outdo each other in deeds of romantic generosity. Those who excelled in the magnificence of their bounty were crowned with wreaths, as if they had conquered at the head of an army. The liberality of Hatim was proverbial, and has immortalized the tribe of Tai. The suppliant he never dismissed from his tent unrelieved. Often were forty camels roasted at a single feast; and, in a season of extreme scarcity, he killed the only horse he possessed,—so valuable, that the Roman emperor had sent an embassy on purpose to procure it. Hatim's benevolence was hereditary; his father rejoiced when he had emptied his folds to feed the hungry, and his mother was interdicted from giving alms for a whole year, lest her prodigality should reduce the family to beggary. He himself distributed the greater part of his flocks among a troop of needy poets on their way to the court of Hira. His beneficence was as unwearied as it was extensive. On the longest and darkest nights he would leave his bed, if some hapless pilgrim required shelter; and, wrapped in his cloak, procure with his own hands a light from some neighbouring tent. Not satisfied with kindling his "fires on the mountains," he would send forth his dog; that, by its barking, strangers might know where to find a place of rest. His memory was revered over all Arabia; and a female captive taken in battle regained her liberty when she pronounced herself to be the daughter of Hatim Tai.*

The fondness of the ancient Arabs for poetry and oratory was excessive; next to the practice of hospitality and expertness in the use of arms and horsemanship, these were the accomplishments on which they chiefly valued themselves. The elo-

* Sale relates a contest that had arisen among the citizens of Mecca, to ascertain which of them was the most generous. The prize was awarded to Abraha, a blind and aged man, who gave away in charity the two slaves that led him,—all the wealth he had,—and groped his way through the city with his hands along the walls.—*Prelim. Diss.* sect. i. Caab and Hassan were celebrated for their liberality.—*D'Herbelot, Biblioth. Orient.* "The Adventures of Hatim Tai," from the Persian, by D. Forbes, is among the works recently published by the Oriental Translation Committee.

quence of Koss was as famous and as flattering a compliment as the liberality of Hatim. Their orations were of two sorts, metrical or prosaic. The former was most esteemed and most cultivated. It was the remark of Abu Teman, who compiled the Hamasa, a collection of old Arabic epigrams, odes, and elegies, that "fine sentiments delivered in prose were like gems scattered at random; but when confined in poetical measure, they resembled bracelets and strings of pearls."* The roving hordes of the desert, living amid the solitary grandeur of nature, were more remarkable for the exercise and admiration of these intellectual endowments than their civilized brethren. Their principal occasions of rejoicing were the birth of a boy, the foaling of a mare, the arrival of a guest, and the rise of a poet. Next, if not equal, to a warrior and a fine horse, this was the noblest possession a tribe could boast of. The genius and merits of the youthful bard were hailed with universal applause. The first-fruits of his muse were commemorated by a solemn banquet, where a chorus of women with musical instruments sang in the presence of their sons and husbands the happy fortune of his tribe. The neighbours flocked to congratulate his family that a champion had appeared to vindicate their rights; that a herald had raised his voice to record their exploits, and recommend their virtues to posterity.

The greatest attention was paid to the cultivation of this divine art. Assemblies of different kinds

* The orientals have always been fond of imagery and metaphor. When an orator speaks, he begins "to weigh his stored pearls in the scales of delivery;" or, "lifting his head from the collar of reflection, he removes the talisman of silence from the treasure of speech, and scatters brilliant gems and princely pearls in his mirth-exciting delivery." If a warrior is slain in battle, "the bird of life has fled from the nest of his brain." A lover says of his mistress, "that the bird of his soul has become a captive in the net of her glossy ringlets."

were held, where rival poets and orators disputed the palm, and took their rank in public opinion accordingly. Each tribe had its annual convention, where its honour was defended, and its deeds celebrated. There were also Panegyrical Meetings, where the actions of their heroes or the munificence of their chiefs were sung, and their respective merits publicly rewarded. In this manner the distinctions of their genealogies were preserved; and the renown of their ancestors, as well as the rights of their families, were transmitted from one generation to another. The most celebrated of these literary convocations was that which took place every year at the fair of Ocadh, near Taïf. Here thirty days were employed, not merely in the exchange of merchandise, but in the nobler display of rival talents. In loud and impassioned strains the contending poets addressed the multitude by turns, extolling the superior glory of their own tribe, recounting the names of their eminent warriors, and challenging their opponents to produce their equals. From the fierce spirit of the Bedouins, and the well-known influence of songs over the martial virtues of a barbarous people, it may readily be imagined that these intellectual battles generally ended in bloodshed.

It was to allay the jealousies and feuds produced by this ancient custom that Mohammed, by a precept in the Koran, expressly abolished it. To conquer in this literary arena was the highest ambition of the bard. The victorious compositions were inscribed in characters of gold upon Egyptian paper, and hung up for public inspection in the temple at Mecca. Of these successful performances seven have been preserved, considered as the finest that were ever written; for the translation of which we are indebted to that distinguished orientalist Sir William Jones. They are known by the name of Modhahebat or the Golden, and Moallakat or the Suspended. Their authors were Amriolkais, Tarafa,

Zohair, Lebeid, Antar, Amru, and Hareth.* As poets in all countries are admirers of beauty, and reckoned peculiarly susceptible of its charms, we find in the Moallakat a greater proportion of love than of any other passion; but this is often mixed with descriptions of local scenery, and sketches of heroes and battles. Sometimes it happened that the young men of one tribe admired the damsels of another, and as their tents were frequently removed on a sudden, the enamoured pair were apt to be separated in the progress of their courtship. Hence the complaints of the unfortunate lover, and his resolution to visit his mistress, form the principal theme of these Arabic poems. He beholds with dismay the empty station where her tent was pitched; and gazes on the silent ruins of the place,—"the black stones on which her caldrons used to boil." He dwells on her beauties and her favours, comparing her to a wanton fawn that sports among the aromatic shrubs. He follows her track through the wilderness "by the locks of stained wool that fall from her camel; until the towers of Yemama appear in the distance,

* Sir W. Jones's Works, vol. x. Antar is the author and the hero of the famous Bedouin romance that bears his name. He flourished about the time Mohammed was born. This work, the genuineness and antiquity of which are incontestable, furnishes a curious picture of the manners and customs of the Scenite Arabs; their descent from Ishmael,—their religion,—mode of warfare,—battles, armour, chivalry, &c. It deals a little in the marvellous. When mounted on his mare Gabrah, our valiant hero kills with his lance 800 men in a single action. One warrior is a mass of solid bone, impervious to every sword but Antar's, which, though not enchanted, had the supernatural virtues ascribed to magic weapons. This romance was first committed to writing in the time of Haroun al Raschid, and still continues to be the principal source whence oriental story-tellers draw their most popular tales. It has been translated into English by Mr. Terrick Hamilton. Antar is the Hercules of Arabia. Burckhardt mentions that when he read portions of it to the Bedouins, they were in ecstasies of delight but at the same time so enraged at his erroneous pronunciation that they tore the sheets out of his hands.

exalted above the plains, and shining like bright sabres unsheathed in the hands of heroes." He describes the storms and dangers he has encountered during this journey; not forgetting to praise the horse or the swift camel on which he rides.

These ancient poems contain many curious traits characteristic of pastoral manners, as well as of the bloody feuds that raged between hostile tribes. The songs of Amru and Hareth relate to the deadly contests of Becr and Taglab; while those of Antar and Zohair celebrate the famous War of the Racehorses between the tribes of Aus and Dobian. The poem of Lebeid has been compared to the Alexis of Virgil. The bard eulogizes the charms, but inveighs against the unkindness of his fair Novara; he then tries to propitiate her, by recounting his own riches and accomplishments,—his liberality and valour,—his noble birth and the high renown of his ancestors. These early effusions of the Arabian muse were the only archives of their nation,—the encyclopædia of their literature,—where their whole stock of useful and entertaining knowledge was treasured up.

To the advantages of a genius for poetry, a lively fancy, and a luxuriance of imagery, the Arabs added that of a copious, flexible, and expressive language. It was derived from the same root with the Hebrew, Syriac, and Chaldee; but was far more rich in its vocabulary. Grammarians have remarked that two principal dialects were in use before the time of Mohammed; the Hamyaric, and that spoken by the Koreish. The former is supposed to nave borne a strong affinity to the Ethiopic; and was so little understood by the northern tribes, that an Arab of Hejaz, we are told, on being directed by the king of Yemen to sit down, threw himself over the parapet, because the word *theb*, in the Koreish, signified, not to *sit*, but to *leap* down. The dialect of the Koreish, being that in which the Koran was originally written, has been always regarded as the classical, or rather the sacred, language of the Moslems.

The extreme copiousness and harmony of the Arabic, enriched by literary compositions and commercial intercourse with different nations, has, from the most remote antiquity, been eulogized by the natives. They assure us that no man uninspired can be a perfect master of it in its utmost extent. That it far outstrips European tongues in this respect, we may be satisfied from the fact that the mere names of a single object, with their explanation, will sometimes fill a considerable volume. The Arabs have two hundred words denoting a serpent, five hundred signifying a lion, and above a thousand different expressions for a sword. Whole treatises have been devoted to the interpretation of these words. Firouzabad, the Johnson of Arabia, the compiler of the great lexicon called the Ocean (Al Kamus), relates, that in his description of the nature and advantages of honey, he has enumerated and explained eighty different names, though there were various others by which it might have been expressed.

This vast accumulation of epithets was the necessary result of their habits and circumstances. The sphere of their observation was limited; and the very paucity of the objects with which they were conversant tended to multiply their expressions. The face of nature, in its rugged and wild monotony, was studied in the desert with a minuteness of which we can scarcely form any conception. To the eye of a Bedouin the aspect of the earth and the sky was infinitely diversified. To his vivid imagination no two clouds were ever alike. The tempest of spring differed from that of summer and of autumn. Every pace of his camel, each period of the life and pregnancy of that useful animal, had its peculiar name. The office of giving it water was differently expressed, according to the number of days it had endured thirst. Every action, motion and neigh of his horse was distinguished from another by an appropriate term. The stranger never came within his view, but he would read his thoughts, dispositions, and affections, in the air of his countenance, the colour of his lip, or the quivering of his muscles. He had as many words to denote a cloud, a rock, a torrent, or a well, as were the almost endless diversities under which these objects daily presented themselves to his contemplation. Yet this immense nomenclature was confided in a great measure to the tablets of memory, and owed its preservation chiefly to the extemporaneous eloquence of an acute though illiterate people. It was under the tents of the wandering shepherds that the language attained its highest cultivation, and where it was spoken with the utmost purity and elegance. Critics have admitted its remarkable delicacy, its bold and energetic sublimity, adapted equally to the simple pathos of love and elegy, the piquancy of satire, or the loftiest efforts of popular oratory.*

The Arabs believe the greater part of their language has been lost; a conjecture not altogether improbable, when we reflect on the perishable record to which it was originally intrusted, and how late the art of writing became generally practised among them. That the use of letters was known in certain parts of the country many centuries before the Mohammedan era we may infer from the testimony of Job (chap. xix. 23, 24), and from the ancient monuments still extant, said to be in the Ham-

* Grammarians have calculated that the inflections of a single Arabic root amount at least to 300 or 350. Supposing the primitive nouns to be 4000 in number, these multiplied by 300 will yield a product of 1,200,000 words; the forms of which can be determined with as much certainty as if every one of them were actually in use. Perhaps 100,000 vocables may be assumed as the greatest number that has ever been required or employed in any language; allowing this quantity, however, to be doubled, it follows that, in the Arabic tongue, there still exists a million of words that have never yet been called into practice.

yaric character. The Christian Arabs of Gassan and Hira were acquainted with letters before their brethren in Hejaz; and the generally received opinion is, that the latter were without an alphabet until the time of Moramur-ibn-Morra of Anbar, who invented the Cufic, and introduced it at Mecca a short time before the birth of Mohammed. After being used for nearly three hundred years, the Niskhi, a more elegant character, was formed from it by the Vizier Ibn-Mokla of Bagdad; and this the Arabs regard as the groundwork of their present alphabet. Until of late years the antiquity of the Cufic, and the subsequent invention of Ibn-Mokla, have been universally admitted. Recent discoveries, however, have overthrown this theory. Medals, coins, and vases have been found with inscriptions older than the Cufic. Papyri have been examined, written in a character resembling the present, earlier by two centuries than Mokla's alphabet; hence the presumption is, that the Niskhi, instead of being imported from Bagdad in the third century of the Hejira, was in use before the Arabs of Hejaz received the other from Irak. That the arts of reading and writing must have been understood at Mecca even in the times of ignorance, we may infer from the Golden Poems suspended in the temple; but it is impossible that Cufa should then have given its name to an alphabet, because that city was not founded till the reign of Omar, more than half a century after the alleged invention.* The Niskhi underwent several changes, and was at length reduced to its present form by Yakut, secretary to Mostasem, the last caliph of the house of Abbas (A. D. 1250). Various Cufic manuscripts on parchment or vellum are in existence, and this character was common in inscriptions on stone and metal until the 13th and 14th centuries of the Christian

* Asiatic Journal, vol. xxiv

era. Even in the present day it is occasionally used in Africa on public monuments, the large size of the letters being well adapted for such purposes. Niebuhr and Chardin have given several engraved specimens of it, found on mosques and sepulchral stones.*

The Hamyaric was very different from the Cufic, or the alphabet of Moramer, and was not allowed to be taught to the vulgar except by special permission. This fact may account for the statement of Pococke, that when the Koran first made its appearance, not a single person in Yemen was able to read it. On the general diffusion of the new religion this primitive alphabet fell into disuse. The Danish traveller was assured that inscriptions in it were to be found among the ruins of Dhafar, and on the walls of a village between Damar and Sanaa. Though prevented from visiting them, he was shown copies, which he thought resembled the Persepolitan or arrow-headed character. The language of the Koran has in its turn been superseded by new idioms. It may now be regarded as a dead tongue, and is studied in the native schools as the Latin and Greek are in the countries of Virgil and Homer.

In what is properly called learning and philosophy, the ancient Arabs had made little progress. They had some pretensions to astronomy; if, to use the words of Sir William Jones, we can dignify with that appellation the mere amusement of giving names to the stars. Their knowledge on this subject was rather the fruit of long experience than of regular study or scientific rules. Babylon, Thebes, and Memphis had schools where astronomy was cultivated; but the only academy open to the native of Arabia was a clear firmament and a naked plain

* Mills's Hist. of Muham. p. 282, note Chardin's Travels. Nieb. Descrip. Arab. Elmacin, Hist. Saracen. chap. iii. Noble's Arab. Vocab. Introd. Meninski, Thesaur. vol. i. p. 16–24. Adle, Mus. Cufic. Borg. p. 34, &c.

Wandering night and day in the open desert, the stars became an object of curious contemplation. They were the guides of his nocturnal marches, and the symbols from which he inferred changes in the air and weather. It was necessary to know their periodical returns, in order to regulate the labours of the field and the succession of the seasons. Like the Indians, this people applied themselves chiefly to the study of the fixed stars; contrary to the custom of other nations, whose observations were mostly confined to the planets: and hence the difference between the Chaldean and Indian idolatry.

Of the high antiquity of this sort of astronomical knowledge among them we have the most decisive arguments both in sacred and profane history. Job, in the sublime Arabian poem that bears his name, speaks in a manner that evinces how familiar, even in his age, were the names and the appearances of the celestial bodies. What he says of Orion, Arcturus, Mazzaroth, and the Pleiades indicates that the nomenclature and phenomena of the science were then no recent discoveries. The natives of the plains of Shinar, and the Ethiopian negroes under the tropic, have each been recorded as the inventors or first cultivators of astronomy. On this point antiquity is not unanimous. That the Sabæans preceded them in this science is hinted by Diodorus, Plato, and Cicero. Lucian expresses himself more plainly when he says, it came originally from Ethiopia,—a name applied to the country eastward as well as westward of the Red Sea, and which his description would here lead us to identify with the Happy Arabia.*

* In his dramatic story of the Runaway Slaves, Lucian introduces Philosophy, who declares to Jupiter, that from the Brahmins she repaired straight to Ethiopia; thence to the Egyptians, whose poets and prophets she instructed; and then she betook herself to Babylon to teach the Chaldeans. This imaginary journey is perhaps a true delineation of the original route of the arts and sciences.—*Landseer's Sabæan Researches.*

The ethics or moral system of the Arabs inculcated some noble and exalted virtues; but it appears to have had little influence on their personal or social habits. Their private intercourse with each other was gross and indelicate. It was not uncommon for the son to marry the father's wife, or for the same man to espouse two sisters. Strabo seems to intimate that some tribes had a community of wives; but these practices were condemned by the more virtuous. Matrimonial infidelity was merely punished with a slight fine; but if the offender belonged to another clan, death was the award. Among some tribes, especially those of Koreish and Kenda, it was the horrid practice to bury their daughters alive, for fear of their being dishonoured or taken captive, and even from the apprehension of being reduced to poverty by supporting them. The birth of females was therefore reckoned a misfortune, and their death a happy relief. If the father intended to bring up the child, he sent her, clothed in a garment of wool or hair, to tend flocks in the desert; if otherwise, he kept her till six years of age, and then directed her to be perfumed and adorned, that "he might send her to her mother;" after which she was thrown headlong into a pit dug for the purpose. But the shocking crime of infanticide was not peculiar to Arabia; it was so common among the ancients, that it was considered a very extraordinary thing that the Egyptians brought up all their children.

With regard to their proficiency in the elegant or the mechanical arts, we have no certain evidence; what we do know leads us to infer that their knowledge of these was rude and imperfect. If the ornaments and costly furniture that adorned their habitations were really the manufacture of their own country, their skill could not be much inferior to the polished ingenuity of Egypt and India. Arrian speaks of sabres, lances, knives, hatchets, and other hardware, fabricated at Moosa; and he mentions certain stuffs of a tartan or striped texture manufactured in Yemen, but not remarkable either for the quality of the wool or the workmanship. Their common military weapons were the bow and the scimitar, in the handling of which they were unrivalled. The other arms which they used were darts or javelins. Herodotus speaks of their arrows barbed with a kind of stone; and Strabo mentions poles of wood, with their extreme points shod with iron or hardened in the fire. These they employed in their wars and predatory excursions. Two archers frequently on these occasions sat back to back on the same camel, that in case of surprise they might be prepared to attack the enemy in front and rear.

Chronometry, or the division of time, among the Arabs was very imperfect. They reckoned ten principal epochs between the arrival of Ishmael at Mecca and the flight of Mohammed, all founded on some historical event, such as the building of the Kaaba, the Flood of El Arem, the War of the Elephant, &c. Different tribes adopted different eras, which tended to darken and perplex their computations. Like most other ancient nations, they divided the year into twelve months, and reckoned by weeks of seven days. They used the lunar calculation; intercalating certain months—one every third year—to make the course of the moon agree with that of the sun, and prevent the seasons from changing their natural order. This arrangement was condemned in the Koran as a remnant of idolatry; and the Arabs now begin their months with the appearance of the new moon, without any intercalation to fix the commencement of the year to the same point; so that the seasons gradually vary their position, and in course of sixteen years winter shifts into the place of summer. As each month begins with the new moon, a cloudy sky will sometimes make the difference of a day's reckoning in two parts of the same country. Their months consist of twenty-nine and thirty days alternately; so that the entire year contains only 354 days. Eleven times in thirty years one day is added to the last month, which makes it amount to 355 days. About the end of the fourth century of our era, the old names of the months and the days of the week were changed by Kelab, an ancestor of Mohammed; and after the Koreish had gained the political ascendency in Arabia, they were adopted by all the other tribes.

TABLE OF THE ARABIAN MONTHS AND WEEKS.

Months.		Months.		Days of the Week.		
	Days.		Days.		Old Names.	Modern.
January	31	Shaban	29	Sunday	Bawal	Yom-ahad.
February	28	Ramadan	30	Monday	Bahun	Yom-thena.
March	31	Shawal	29	Tuesday	Jebar	Yom-tulta.
April	30	Dulkaada	30	Wednesday	Dabar	Yom-arba.
May	31	Dulhajja	29	Thursday	Femunes	Yom-hamsa.
June	30	Moharram	30	Friday	Aruba	Juma.
July	31	Saphar	29	Saturday	Shiyar	Sabat.
August	31	Rebiah I.	30			
September	30	Rebiah II.	29			
October	31	Jomadi I.	30			
November	30	Jomadi II.	29			
December	31	Rajeb	30			
	365		354			

Note.—To reduce the Christian era to the Mohammedan, take 622 from the current year, multiply by 1.0307, cut off four decimals, and add .46; the sum will be the year and decimal of the day. This rule is calculated for the Old Style; but the correct answer will be found by deducting the difference between the styles. The present year, 1833, corresponds to 1248 of the Mohammedans; the month Ramadan began on January 22.

Like all ignorant people, the Arabs were addicted to superstition. A favourite study among them was oneiromancy, or the interpretation of dreams. This art and various other kinds of divination they had in common with the Jews, Chaldeans, Egyptians, and other nations. The country swarmed with magicians, sorcerers, soothsayers, and astrologers. These "wise men" held different ranks, according to the depth of their knowledge. The cohen or diviner appears to have been the chief priest of the whole sect. It was his office to know things that had never happened,—to predict all contingent events, —to interpret dreams, and even tell and explain visions which the dreamer had forgotten,—to be questioned and give answers on any subject a stranger wished to know,—to tell what was passing in distant places,—to give news of absent persons, and discover goods that had been lost or stolen. Most of these soothsayers pretended they were inspired by a peri or fairy; for these little green genii, so popularly known in this country, were natives of the East, and may be classed among the poetical fictions which the Saracens brought to Europe. Kings applied to these wizards in their perplexities. When the last of the Tobbaas had a vision, which he could neither relate nor comprehend, he assembled four hundred of them, at the head of whom were the famous Shak and Satih. They related and interpreted to him the particulars of his dream, which portended that a black race from Ethiopia would seize his kingdom, and abolish the Jewish religion.

To augury, and other popular superstitions, the Arabs were extremely devoted. The flight of a bird, or the particular motion of an animal, was sufficient to suspend the most important journey. If the former flew to the right it was a lucky omen, if to the left the traveller returned home. The most ordinary events in life,—justice, medicine, courtship, matrimony,—were regulated by the imaginary influence of certain spells. A hare's foot suspended round the neck, or in any part of the house, was a charm against the witchery of an evil eye, and protected the family from all sorts of male demons, whether from the woods or the deserts, that haunt the abodes of men. The tooth of a fox or a she-cat, or the red gelatine exuded from a prickly shrub (*Spina Egyptia*), was worn as an amulet, to prevent blindness or other malignant influence of female demons. If the youth of either sex wished to have a fine and healthy set of teeth, they had only to take a cast tooth, and holding it between the forefinger and thumb, throw it towards the sun, with a brief petition to the gods to substitute a better. When pustules appeared on the lips or face, the person afflicted went a-begging from house to house, crying *aljeba*, with a sieve on his head, into which were thrown crusts or fragments of meat; and on these being distributed among the dogs the disease was supposed to vanish. A traveller, if he dreaded the plague, before entering a city, brayed at the gate ten times in imitation of an ass; and this was the charm he used against infection. To cure the sting of a scorpion, the patient was wrapped for seven days in a woman's garment, with her bracelets and earrings hung above him; during which time he was kept awake by loud and incessant noise. A suspicious husband, before setting out on a journey, used to cast a knot secretly on a stalk of broom; if he found it tied on his return, he inferred his wife had been faithful; if loose, that he had been dishonoured. On the death of a husband, the widow left the tent, clothed herself in a mean dress, using neither aromatics nor ornaments for a whole year. She was permitted to touch no animal but a dead ass, a sheep, or a fowl. At the end of the year her weeds of sorrow were thrown aside, and the use of her ointments and perfumes resumed. It was the practice to leave the favourite camel to starve at the grave of its master, with its head fastened to its tail, and covered with his cloak, that the defunct at the resurrection might not be obliged to travel on foot in the other world. If the animal made its escape it was held sacred, and allowed water and pasture free. The women never mourned the slain until their death was avenged, but their lamentations began the moment expiation was made. Before due vengeance was obtained, an owl or some ominous bird (named Hamah) was supposed to sit and cry for "drink" on the grave, until its thirst was allayed by the blood of the manslayer.

Judgment by fire was an ordeal in common use, and by its decision private quarrels were frequently adjusted. Dice, lots, chess, and other games of chance were converted into superstitious practices. The sortilege, or gambling by arrows, called *almaisar*, was much in use. Its object was to gain money, by throwing on a few the risk of a certain enterprise or amusement in which several were embarked. For example, a young camel was purchased in common, killed, and divided into ten or twenty parts. Seven persons were appointed to cast lots for them; and eleven arrows provided, without heads or feathers, seven of which were numerically marked, and four plain. These were thrown promiscuously into a bag, and drawn by an indifferent person. Those who had the notched arrows won shares according to the number of their lot; while the blanks were entitled to no part, and obliged to bear the whole expense.

Another popular superstition was the *azlam*, or divination by arrows; those used for the purpose being kept in the temple of some idol, in whose presence they were consulted. The art was thus performed: three arrows were enclosed in a vessel; on the first was inscribed "God command me;" on the second, "God forbid me;" the third was plain. If the first was drawn out, the suppliant prosecuted his design; if the second, he deferred it for a year; if the third, he drew again until he received an answer,—not forgetting to repeat his present to the idol or the priest each time. No affair of importance was undertaken—be it a journey, a marriage, a battle, or a foray—without the advice of these sacred implements. Matters of dispute, such as the division of property or plunder, were settled by an appeal to them. The ancient Greeks practised this sort of divination, as did the Chaldeans; for we learn from Ezekiel (chap. xxi. 21), that the king of Babylon, in marching against Jerusalem, "stood at the parting of the way to use divination, making his arrows bright" (or, as Jerome explains it, mixing and shaking them together), that he might know which city first to attack.

The pagan Arabs were grossly idolatrous. Though assuming a variety of forms, the essential basis of their religion was Sabaism, or star-worship,—the primitive superstition of most oriental nations. The number and beauty of the heavenly luminaries,—the silent regularity of their motions,—the sun rejoicing to run his race,—and the moon walking in brightness,—were all calculated to impress the vulgar mind with the idea of a superintending and eternal power. From viewing them as the visible types of a divinity, man, in the simple infancy of his reason, believed them to be endowed with instincts like his own; animated with his understanding, and subject to his passions.

But when to this childish error was added the general persuasion of their real or imaginary influence over the productions of the earth and the fortunes of its inhabitants, the transition from curiosity to adoration was natural and easy. When the husbandman observed the growth of seeds and plants to maintain a constant and invariable sympathy with the phenomena of the heavens, and vegetation flourishing and disappearing with the rising and setting of certain planets, or the same group of stars; and when the shepherd remarked the increase of his flocks, and the genial moisture that enriched his pastures, harmonizing with the periodical return of the celestial bodies,—they learned, as it were mechanically, to associate in their minds the operations of the one with the constant recurrence of the other; and even applied to the heavenly hosts the very names of the terrestrial objects to which they seemed linked by some mysterious affinity. The Bull and the Ox were the stars that indicated the season for ploughing and preparing the soil; the Ram, the Lamb, and the Goat were the signs under which these valuable animals brought forth their young. The Lion and the Dog were venerated for the same cause; the group of the Crab measured the boundary of the tropic, when the sun, having reached the limit of his southern journey, turned backward on his annual path. The Scorpion was the terrible harbinger of the burning and poisonous winds; and the Balance marked the annual equinox, when the day and night are of an equal length, resembling the equilibrium of that instrument. Hence the origin and the corruption of stellar worship, the most obvious and innocent of all idolatries.

The adoration of the stars was natural to the lively and credulous Arabs; but the strange diversity of their idols seems a contradiction to the uniformity of their habits in other respects. Some faint traces of the patriarchal religion still lingered among them, for they were not ignorant of the unity and perfections of the Deity. This Supreme Being they called *Allah Taalah*, or the Most High God; but their fantastic creed embraced a variety of subordinate divinities. They had seven celebrated temples dedicated to the seven planets: that at Mecca was consecrated to Zohal or Saturn; the Beit Ghamadan, at Sanaa, was built in honour of Zoharah or the planet Venus. The Hamyarites chiefly worshipped the sun: they had a famous edifice at Aden; and the enthusiasm of the worshipper must have burned with redoubled ardour when he saw from the rocky precipice that glorious luminary rising like an orb of fire from the bosom of the Indian Ocean. Some tribes exclusively revered the moon, some the dog-star, others the planets Jupiter and Mercury. The religious festivities of these divinities were regulated by the sacred constellations, and generally celebrated at the terms of the equinox and the sol stice: as we learn from Nonnosus, who, in the time of Justinian, passed through that country on an embassy to Ethiopia.

It was the belief that the stars were the dispensers of weather, which led to the idea of their being inhabited by angels, or intelligences of an intermediate nature between man and the Supreme Being; hence the Arabs paid them divine honours, because of the alleged benefits they procured through their intercession. Of these sidereal divinities the Koran mentions three that were worshipped under female names,—Al Lattah, Al Uzzah, and Manah. The first had a temple at Naklah, near Taïf; the second was adored by the Koreish; and the third by the tribes of Hodhail and Khozaah. Manah is described as a rude unshapely block of stone, of a black colour, four feet high, and two broad, fixed on a golden pedestal. Five other idols are specified in the Koran: Wadd, worshipped under the human figure by the tribe of Kelb; Sawah, a female deity, adored by the tribe of Hamadan; Yauk bore the resemblance of a horse; Nasr that of an eagle; and Yaghuth, a popular deity in Yemen, that of a lion. Hobal was one of the most famous of these idols. This statue, brought from Belka in Syria, was the figure of a man made of red agate, and placed on the top of the Kaaba, near the images of Abraham and Ishmael. Having by some accident lost the hand that held the divining arrows, the Koreish substituted one of gold. Around him stood a swarm of inferior deities, which had accumulated to the number of 360; so that at Mecca alone the Arabs might approach a fresh object of devotion every day in the year. The Hanifites had a lump of dough for their god; which, in cases of extreme famine, they used, as the Egyptians did their leeks and garlic, "at once for worship and for food." The images of men, women, beasts and birds which crowded the

Arabian pantheon were almost innumerable. Every head of a family, every householder, had his tutelar god, which received his last adieus when he went abroad, and his first salutations on returning home They named their children after their favourite idols, and gloried in being reckoned their servants and votaries.*

Some tribes, from their frequent intercourse with Persia, had imbibed the religion of the Magi or fire-worshippers, while others had become converts to Judaism. That the doctrines of the cross were early received in Arabia will admit of no dispute. We know that Christianity ranked certain Arabs among its first converts, some of them being present on the day of Pentecost. It is the universal belief in the Eastern churches, that the apostle Thomas preached in Arabia Felix and the island of Socotra (A. D. 50), on his way to India, where he suffered martyrdom. St. Paul himself resided in the kingdom of Gassan (Galat. i. 17); and it is highly probable that the Arabian merchants who visited the fairs of Bosra and Damascus must have heard him, and perhaps were converted by his discourses. The dispersion of the early Christians would doubtless scatter the seeds of truth over various regions; and we may suppose that many victories over ignorance and error would be achieved by the translations of the Sacred Books, as well as by the able works that

* Sale, Prelim. Diss. sect. i. Pococke, Not. in Specim. Arab. Hist. Univ. Hist. vol. xviii. b. iv. chap. 21. Selden de Diis Syriis, tom. iii. Ὁ δε ρηθεις λιθος παλαι προσεκυνουν δι Ισμαελειται is the remark of an old author on the litholatry of the Arabs Forster's Mahom. Unveiled, vol. ii. p. 408. That the altars of the pagan Arabs were stained with human gore we learn from Porphyry, who says, that in the third century a boy was annually sacrificed by the tribe of Daumath. The horrid practice of Παιδοθυσια and Ανδροθυσια is attested by Procopius, Bell. Persic. lib. viii. c. 28; Evagrius, lib. vi. c. 21; and Pococke, p. 72, 86 Hyde. Hist. Relig. Vet. Pers. p. 129.

were written to expound their doctrines, and vindicate their authority.

Much was done in this way by the zeal and labours of Origen, a presbyter of Alexandria; who, in point of erudition, was the greatest luminary of the church in his time. We learn from Eusebius that, on the invitation of an Arab prince, he repaired to that country, and succeeded in converting a certain tribe of Bedouins to the Christian faith. His services in behalf of the disciples in Arabia were likewise signalized in the field of controversy. A sect of philosophers had arisen there, the followers of a master whose obscurity has concealed him from the knowledge of succeeding ages, who denied the immortality of the soul. Origen was summoned to oppose these heretics: in a full council he disputed their tenets with such ability and success that they abandoned their errors and returned to the orthodox belief. Among other conversions achieved by the arguments of this Father, we find that of Beryllus, bishop of Bosra, who had espoused the doctrines of the Monophysites. Unable to resist the eloquent reasoning of his antagonist, he candidly yielded the cause, and again avowed himself a believer in the pre-existence of Christ's divine nature.*

In the third and fourth centuries, Arabia became the common asylum for numerous victims of persecution under the Roman emperors, chiefly of the Jacobite or Monophysite persuasion. At that time the principal Christian tribes were those of Hamyar, Rabiah, Taglab, Bahra, Tanuch, and part of Tai and Khodaah. Nearly all the inhabitants of Nejeran were Christians in the reign of Dunowas. The Gassanites had been converted at a very early period. Abulfeda states that many churches were planted in Hira. Mondar declared war against the Emperor Justinian, for oppressing the Jacobites on account of their supposed heterodoxy; so we may

* Mosheim's Church Hist. vol. i. chap. v

presume that Christianity must have been the popular or established religion there. Procopius has a remark which corroborates this supposition; for he asserts, that "the disciples of Christ had filled the provinces of Arabia with the churches of God."

Whether or not the natives possessed a version of the Scriptures in their own tongue is a point undetermined by direct evidence. No translation is known to have existed; but Gibbon infers, from the practice of the synagogue in expounding the lessons in the vernacular dialect of the country, and from the assertions of the Fathers in the fifth century, that the Scriptures were translated into "all the barbaric languages," that the Christian Arabs possessed copies of the Gospel in their own idiom. Another and a stronger presumption in favour of this inference is, that a number of episcopal sees were established in different parts of Arabia; and we can hardly imagine these venerable prelates to have been entirely without translations of the religious books they were appointed to teach. Nor is it likely that Origen, who disputed with the schismatical doctors, and did so much to transcribe the Gospel into other tongues, would have neglected this mode of converting the idolatrous Arabs. In the northern provinces numerous dioceses are mentioned; Suez, Sinai, Feiran, Petra, Akaba or Ailah, Bosra, and other places, could boast of episcopal dignitaries at a very early period of Christianity; and some of them are mentioned as taking part in the discussions of the general councils. The Jacobites in Irak had two bishops, both subject to the Mafrian or metropolitan of the East; one of them resided at Akula (near Bagdad), the other, who was styled Bishop of the Scenite Arabs, had his seat at Hira. The Nestorians in that district had but one bishop, who was immediately subject to their own patriarch. The tribe of Tai had likewise their prelate; but the extent of his jurisdiction has not been very clearly defined. In Arabia Felix we find three bishopricks

Nejeran, Dhafar, and Aden; and it is probable that Sanaa, Damar, and other towns under the Abyssinian kings, enjoyed a similar distinction. About the time of Dunowas we find Gregantius, bishop of Dhafar, celebrated as a controversialist; though his exploits in converting the Hamyaric Jews must be rejected as an extravagant fable.*

As Arabia had been a kind of sanctuary for the proscribed and persecuted exiles of all sects and denominations, we may naturally suppose that its churches were overrun with the prevailing errors and corruptions which unhappily were soon grafted on the pure and simple doctrines of the apostles. The facility with which the Arabs embraced the absurdities of paganism seems to have disposed them to a like readiness in falling in with the Christian heresies. The principles of the Ebionites and Nazarenes, who denied the divinity of Christ,—of the Nestorians, who taught that he had not only two natures, but two persons,—and of the Collyridians, who paid divine honours to the Virgin Mary, were widely propagated among them. The latter were extremely popular among the female sex, who judged it necessary to appease the anger or court the favour and protection of their "blessed goddess," by libations, sacrifices, and oblations of cakes (collyridæ).†

At that period the endless schisms and heresies that rent asunder the entire fabric of Christianity may be said to have been in their zenith. The Millenarian, with his thousand years of celestial felicity

* Sale, Prelim. Diss. sect. i. p. 30. Pococke, p. 136.

† "Ferax hereseon Arabia" is the expression of St. Augustine. Hottinger, Hist. Orient. p. 212–231. Olaus Celsius, Hist. Ling. Arab. Roderic Ximenes (Archbishop of Toledo), Hist. Arab. c. i. The sins of the Eastern countries, and chiefly their damnable heresies, hastened God's judgments upon them. In these western parts, heresies, like an angle, caught single persons, which in Asia, like a drag-net, took whole provinces.—*Fuller's Holy Warre*, b. i. c. 6. *Epiphan. de Hæres* lib. iii *Eutych Annal*

on earth; the Manichæan, with his incongruous mixture of Persian philosophy; Arians, Eutychians, Gnostics, Montanists, Marcianites, Sabellians, Valentinians, and a host of obscurer sects, all rose up in the theological arena, to foment new divisions, and perplex religion with trivial and unintelligible distinctions. Each of these had their leaders and abetters, whose names gave a sanction to the wildest reveries that human imagination could invent. Of their deluded followers, some macerated their bodies with hunger and thirst; some tore their flesh with scourges of whipcord; and others, tired of terrestrial vanities, shut themselves up in dens and holes of the earth, leading a life more worthy of a savage animal than a rational being. To this universal degeneracy of manners and opinions were added the vices that degraded the character of the clergy. The primitive examples of peace, love, and charity, of singleness of heart and disinterested zeal, had vanished amid the struggles of jarring factions and ambitious prelates. The infatuated disputants contended with implacable fury about points the determination of which lay beyond the reach of human intellect. It was the remark of a heathen writer, not more severe than true, concerning the Christians of the fourth century, "that their animosity towards each other exceeded the ferocity of the beasts against man."* The quarrels of rival dignitaries cast a reproach on the faith of which they were the unworthy defenders. So keenly were the supreme honours of the church contested, that episcopal elections became scenes of bribery, violence, and murder. Damasus and Ursicinus, at Rome, in the year 366, carried their priestly strife to such an extreme, that the city was given up for the time to anarchy and massacre, not fewer than 137 persons being found killed in one day in the church of Sicininus. Such was the lamentable state

* Ammian. Marcellin. lib. xxi. xxvii.

of religion and morals,—of heretical divisions and clerical degeneracy, which paved the way for the downfall of the Eastern Church; and such were the favourable opportunities held out to the daring fanaticism of the Arabian Prophet for establishing that gigantic superstition, which so soon threw its baleful shadow over the first conquests of the apostles, and the fairest provinces of Christianity

---

CHAPTER VI.

LIFE OF MOHAMMED.

Contradictory Views of Mohammed's Life and Character—His Birth and Education—Visits Syria as a Merchant—Marries Kadijah—Affects an austere and retired Life—Proposes to reform Religion—Assumes the Title and Office of the Apostle of God—His first Converts—Announces publicly his prophetic Mission—His unfavourable Reception—His Proselytes increase—Miracle of splitting the Moon—The famous Night Journey to Heaven—His Secret League with the Medinian Converts—The Koreish resolve to put him to Death—His Escape and Flight to Medina—Proclaims a Holy War against the Infidels—Battle of Bedr—Defeat of the Moslems at Ohud—Quarrel with the Jews—Siege of Medina—Expedition to Mecca, and Truce with the Koreish—Siege and Capitulation of Khaibar—Attempt to poison the Prophet—Mohammed sends Letters and Ambassadors to Foreign Courts—Respect shown him by his Followers—Battle of Muta—Capture of Mecca by the Moslems—Demolition of Idols and Images—Battle of Honain—Surrender of Taïf—Expedition to Taouc—Increased Power and Success of Mohammed—His valedictory Pilgrimage to Mecca—His Sickness and Death—His personal Appearance—His private Character and Habits—His Wives and Concubines—His supposed Ignorance of Letters—Concluding Reflections.

The life of Mohammed, and the peculiar institutions of which he was the author, have been treated at great length, and by an infinite number of writers; some of whom have spoken of them in a tone of bitter hostility, others in a style of panegyric, that destroys all confidence in their veracity. The narratives of the Mussulmans themselves, from whom it was natural to expect the most authentic and satisfactory accounts, as being the collectors, if not the actual witnesses, of the circumstances they relate, are unsafe guides. These writers had a deep interest in the fame of their prophet, which has tinged their histories with extreme partiality, and must greatly depreciate the value of their testimony. The Christian annalists we can hardly suppose to be more trustworthy in this particular than their Mohammedan opponents. Hating both the creed and the apostle of the infidels, it is not likely they would give a fair representation even of the truth; or that they would spread any reports but such as were to his prejudice, and which might tend to bring his impious forgeries into derision.

Though much uncertainty on this subject has been removed by our increased acquaintance with the literature of the East, and a more candid spirit of investigation introduced, there still remains considerable obscurity respecting the personal history of Mohammed. The narratives of his life are broken and disjointed, resting sometimes on equivocal evidence, and very often enveloped in a thick cloud of supernatural wonders, that makes it difficult to separate between earth and heaven, or discriminate the exact bounds of truth and fiction.* To dignify

* The authors who have written Lives of Mohammed it would be tedious to enumerate. The best Arabic biography yet discovered is that by Abulfeda, which was translated into Latin in 1723, and illustrated with copious notes by John Gagnier, Professor of Arabic at Oxford. This work, for a Mussulman, is candid and judicious. Al Beidawi, Shahrestani, Al Jannabi Nuvairi, Mircond, and most of the other oriental historians, are full of legends, and not worth noticing here: they have been consulted and copiously used by D'Herbelot and the authors of the Universal History (Mod. Part, vol. i.). The Lives of Mohammed, not mere translations, but compiled from various

these reveries with the name of history would be an insult to common sense; while to discard them entirely would have shorn, in the opinion of all true Mussulmans, the character of their prophet of its greatest glory. We have deemed it the more proper course to record the statements of these authors as we found them; satisfied that the broad line of separation, between the probable and the miraculous, will of itself point out to the reader what he ought to reject and what to believe. It falls not to the lot of ordinary mortals, for instance, to be exempted from original sin, to converse familiarly with angels, to split the moon, or make a personal excursion through the seven heavens; yet these and other marvellous exploits are gravely ascribed to the Arabian apostle. Such ridiculous extravagance stands self-refuted, and requires no antidote or contradiction.

Mohammed (or Mahomet, as he is improperly called) was born at Mecca; but the precise date of his birth has been disputed. The computation most generally approved has fixed it in the 569th year of our chronology.* The calumny of his early Chris-

authors, are innumerable That by Dean Prideaux, published in 1697, has been long popular; it is learned but dull, compiled from suspicious authorities, and tainted with prejudice. The one by the Count de Boulainvilliers, which appeared in 1730, is deserving of no credit, either for its sentiments or its consistency with fact. It is a preliminary essay or romance rather than a history, being a mere fragment, and bringing the life of the Prophet only down to the fifth year of his mission. The learned Abbé Maracci prefixed a life, full of bitter invective, to his Translation and Refutation of the Koran (in 1698). Gagnier compiled a biography, in 2 vols., from the Koran and the best Arabic authors, in 1732. He is impartial; but he writes like a Mussulman,—recording facts and fables, miracles and visions, with the same imperturbable solemnity, and without a single remark. That prefixed to Savary's Translation of the Koran is an excellent abridgment of the Prophet's Life.

* Or in 571, the lunar reckoning of the Arabs making a difference of more than two years. Elmacin and Abulfarage adopt

.ian adversaries, who sought to debase him into a man of low plebeian origin, has long been exploded; for whatever uncertainty may rest on the first gradations of his pedigree, he could, without doubt, number among his ancestors in a direct line many generations of pure and genuine nobility.

The Arabs glory in the number of their children; it is the highest honour and ambition of their wives.* In this respect Providence had richly crowned the domestic happiness of Abdolmotalleb; he was the father of six daughters and thirteen sons. Abdallah was a younger son,—the best beloved of his father,—the most beautiful and amiable of the Arabian youth. His personal charms are said to have exposed him to many temptations, similar to what the virtuous Hebrew captive encountered in the service of Potiphar. The maidens of the Koreish pined in secret admiration, and eagerly courted his alliance. About his twenty-fifth year he was united to Amina, daughter of Waheb, chief of the Zahrites, a family of princely rank, being also descended from the Koreish. She was famed for her wisdom and beauty, and had been sought in marriage by the wealthiest of her kinsmen. The consummation of their nuptials was fatal to many languishing hopes; for the tradition of Ibn Abbas assures us, that on the same night 200 virgins expired of jealousy and despair. There was not a soothsayer, we are told, or prophetess, in all Arabia, but had intimation of the auspicious event; and not

571; Reiske, 572. The Benedictine monks (Art. de Verif. des Dates) fix it on the 10th of November, 570. Prideaux and Hottinger, on the authority of Arabian writers, remove it to May, 571. Gagnier makes it 569 of the vulgar era, or 578 from the birth of Christ.

* "It was the custom, that when a woman brought forth ten male children, she should be called Munejeba (or the ennobled), and her name be published among the Arabs; and they used to say that the wife of such a one is ennobled.' —*Romance of Antar*, p. 2'

a bride but sighed to be the mother of a male child. The sole fruit of this union was Mohammed. His father being then absent with the caravans in Syria, to purchase a supply of provisions (for it was a time of scarcity), died at Medina on his return; and whether this event took place before or two months after the birth (according to Abulfeda), it appears certain he never saw this wonderful infant. Cut off in the flower of life, his orphan child was left in the cradle to the care of a widowed mother. Adversity seemed to be his only patrimony; for Abdallah, though of royal lineage, was poor; and in the division of his inheritance, the prophet's share was but five camels, and an Ethiopian slave, named Baraca. On the seventh day, the venerable Abdolmotalleb made a splendid entertainment to the grandees of his tribe; and on this joyous occasion the infant received the name of Mohammed, contrary to the remonstrances of the Koreish, who would have preferred a name that was hereditary in the family.*

Not content with the narrative of simple facts, the credulous superstition of the Arabs has thrown a halo of wonders round the infancy of their apostle. Though destitute of worldly wealth, his birth was rich in prodigies. Like that of other great men who have astonished the world, it was accompanied by signs in heaven and miracles on earth. The prophetic light that surrounded him served his mother for a lamp, and shone with a brilliancy that illuminated the country as far as Syria: the sacred fire of the Persians, which had burned without interruption for a thousand years, was for ever extinguished; the palace of Khoosroo was rent by an earthquake, and fourteen of its towers levelled with the ground; events that prefigured the failure of the

* This name is the past participle of the verb *Hamad*, and signifies the "praised," or "most glorious." Islam, the religion of Mohammed, and Moslem or Mussulman, come from the same root, *eslam*, which means consecrated or dedicated to God.

royal line of Persia, and the subjugation of that country by the Arabs, after the reign of fourteen kings.

A vast multitude of other fictions and supernatural prognostications, equally extravagant, were carefully collected by his biographers. They were devoutly believed, even during his life, by his credulous followers; hundreds of whom were to be found, who, on their oath, would have attested the reality of them. These solemn wonders were obviously invented and propagated with a view to assimilate the birth of their pretended apostle to that of the Founder of Christianity; many of the Scripture texts prefiguring the one, and the marvellous circumstances accompanying his nativity, as recorded in the apocryphal gospels of St. Thomas and St. Barnabas, being literally appropriated to the other.

The nurture and education of her only child had devolved on Amina; but the custom of the Arabian nobility, and the unwholesome air of Mecca, made it necessary to delegate this maternal task to other hands; and the names of the different nurses that suckled this remarkable infant, the most celebrated of whom was Halima, of the tribe of Saab, have been scrupulously recorded. These facts, however, as well as his speaking in the cradle, and his purification from original sin by an angel, we leave to be studied in the legendary pages of the Moslem biographers.

In his sixth year Mohammed was deprived of his mother. This second calamity threw him entirely on the charity of his grandfather. Within ten years the venerable Abdolmotalleb expired at the age of 110. Abu Taleb became his next protector, who appears to have been the eldest son and successor to his father's authority.* The uncle treated the

* Prideaux calls Abdallah the eldest son, and Boulainvilliers supposes him the youngest. Both are mistaken (vide Sale's

orphan boy with paternal affection. The pure air and hardy nutriment of the desert (the Arabian children went quite naked) had already laid the foundation of a robust constitution; the elements of a rough and scanty education were now supplied by the kindness of his relative.

A blank of five years has been filled up with inventions. We can only conclude that being designed for a mercantile life his instructions were likely to be suitable to his profession. At the age of thirteen he made a commercial journey to Syria in the caravan of his uncle. This expedition is barren of facts, but the void is occupied with imaginary adventures. Certain it is, that here the youthful merchant had an opportunity of signalizing his courage, or rather serving his first campaign in the ranks of his clansmen. But tradition has made this journey still more remarkable by several wonderful indications of his future elevation. It was in the fair of Bosra that he is alleged to have met the celebrated Nestorian monk, Felix or Sergius, surnamed Bahira, who is accused by the Christian writers of afterward assisting him in the contrivance and composition of the Koran. Till his twenty-fifth year nothing further is recorded of his history. Some modern writers, such as the Count Boulainvilliers, during this interval, have schooled him in martial exercises; inured him to hunting and other manly pastimes, and carried him in imaginary voyages over all the East; but these we omit, as they have not the sanction of any Arabian author.

His probity and talents for business introduced him to the acquaintance of Kadijah, a rich widow in Mecca. She was of noble extraction, her father, Kowailed, being of the tribe of the Koreish. En-

Prelimin. Discourse, sect. ii. p. 50). Abulfeda makes Hamza and Al Abbas both older than Abu Taleb, and younger than Abdallah.—*Gagnier*, p. 67.

gaged extensively in traffic, she had realized an independent fortune. To her he was recommended by Abu Taleb as factor; and, as her promises were liberal, he undertook the superintendence of her affairs. Her esteem for him was increased by his fidelity in her service, and gradually ripened into affection. In this new capacity Mohammed made a second journey to Syria, where the interests of his mistress demanded his presence. The sales he effected of the merchandise intrusted to his charge were highly advantageous,—a circumstance credible enough in itself, without resorting to the fiction of the Arabs that his goods drew a double price in all the Syrian markets. Having made the necessary purchases, Mohammed prepared for his return. Kadijah, who had thought his absence long, was transported with joy at his success. Her heart, already half-won by the charms of his person, now burned with a passion she could not control. Far from resisting this honourable attachment, she offered him her hand and her fortune,—a generosity which he accepted with cheerfulness, and remembered till the last day of his life with gratitude.

At the time of his marriage Mohammed was twenty-five years of age, and the wealthy Kadijah was forty. Two husbands, Atik and Abu Halat, she had already laid in their graves; but she enjoyed the reputation of much prudence and an irreproachable life. Their nuptials were celebrated with great festivity, mirth, music, and dancing. But we shall pass without comment the angelic demonstrations of rejoicing on this occasion—how a heavenly voice pronounced a benediction on their union—how the boys and girls of paradise were led out in their bridal garments—how the hills and valleys capered for joy, and precious ointments were diffused over the whole earth. Among the terrestrial guests at this ceremony were Abu Taleb and the chief men of the Koreish. The marriage-contract, the formula of which has been preserved, recites, in the simple style of Arabian manners, the noble lineage of the bridegroom, his beauty, virtue, intelligence, and poverty; the reciprocal loves of the happy pair, and a promise on the part of Abu Taleb to pay the marriage-dowry of his nephew, which consisted of twelve ounces of gold, or (according to Abulfeda) twenty young camels.

Some have imputed a reluctance on the part of the bride's father to this alliance, as dishonourable to his family; while others would degrade the commercial agent into a mere driver of camels, or a menial servant in the household of his mistress. Both suppositions have originated in malice. Kadijah could suffer no degradation in the esteem of her kindred by a match with the grandson of their prince, young, handsome, and affectionate. Neither was there any reproach of servitude or dishonour in the prosecution of commerce. It was a lucrative occupation, in which the noblest and bravest were engaged. Sovereigns reckoned it no indignity to command their own caravans. This office was not merely a simple affair of merchandise,—it was also a military expedition, where they had to defend themselves and their cargo from the robbers of the desert. To conduct an escort of this kind was in truth to superintend the finances of the state, and to maintain the freedom of trade, by repelling the aggressions of troublesome and rapacious enemies. Whether or how long Mohammed continued after his altered circumstances to cultivate his fortune by traffic, is a question that has not been solved; nor is it of the smallest importance. Content with his domestic virtues, Kadijah became the mother of four sons and four daughters. This wealthy alliance restored the heir of Abdallah to the splendour of his ancestors, and to a station of equality with the richest in Mecca.

The total silence of his biographers for fifteen

years has here left a cloud of mystery to hang on the life and actions of Mohammed. We merely learn from Abulfeda, that God had inspired him with a love of solitude and retirement; and that every year, for a month at a time, he withdrew to the cave of Mount Hara, three miles from Mecca, where he devoted himself to fasting, prayer, and meditation. This studied and sanctimonious austerity was no doubt preparatory to his grand design. Retirement is not only the school of genius,—it is the fitting nursery of enthusiasm. The practice has been common in all countries. It is amid the solitudes of groves and grottoes, far removed from the bustle and distractions of the world, that the mind of the visionary finds its proper element: witness the converse of Minos with Jove on Mount Ida, and the nightly visits which Numa received from his fabled nymph Egeria. The oracles or spiritual beings they consult dwell not in cities or crowds, but among the echoes of woods and rocks. The affected seclusion of the Arabian impostor was conformable to that of other enthusiasts. In the solemn obscurity of Hara he laid the foundation of his future greatness; for it was in the silence of that retreat that he meditated the scheme of his religion, perhaps the subjugation of his country.

It has been matter of controversy, whether in these transactions Mohammed ought to be regarded as a cunning knave or the dupe of enthusiasm. The point is scarcely worth the disputing; for no imposture, civil or religious, was ever successful without a mixture of both. Had the Arabian adventurer been the mere dupe of a heated imagination, he might have continued to preach his doctrines with all the fervour of an apostle, among the tribes of the desert or the tents of the pilgrims; but his piety would hardly have dreamed of cutting its way with a sword to a temporal throne. Fanaticism was with him an earlier passion than ambition, and most likely supplied the first materials for the great political structure which he afterward reared on this basis. Instead of religious innovations, had his aim been merely secular aggrandizement, there was much in the condition both of his own and the surrounding nations favourable to his revolutionary projects. No usurper, perhaps, ever enjoyed these advantages to a greater extent. Nor can we suppose that a vigorous and reflecting mind like his, enlarged by travel and observation on mankind, could lack either courage or discernment to turn them to his interest. The political state of the Eastern World was wretched in the extreme. Exhausted with continual wars, and enervated by luxury, it could offer little resistance to any aggressor. Had the Roman empire retained its pristine vigour, the Arabian heresy must have been instantly crushed, or driven to the inaccessible retreats of the mountains. Its hapless founder might have been condemned to the stake by a council of bishops, or carried in chains as a rebel to languish out his days in some dungeon of the Grecian capital. But this mighty power had fallen, under the successors of Constantine, into a state of weakness and decay. The Goths in the west, and the Huns in the east, had overrun its finest provinces, and made the once potent Cesars tributaries to a barbarous conqueror.

But whatever information Mohammed had, or whatever use he designed to make of the advantageous posture of oriental affairs, his grand and earliest object of attention was the idolatry of his countrymen. He did not pretend to introduce a new religion; for that would have alarmed the jealousies of all parties, and combined their discordant opinions into a general opposition. His professed object was merely to restore the only true and primitive faith, such as it had been in the days of the patriarchs and prophets, from Adam to the Messiah. The fundamental doctrine of this ancient worship, which he undertook to purify from the alloy it had unhappily contracted among a frail and degenerate race of men, was the UNITY OF GOD.

A principle thus simple and obvious, which no sect had ever denied, and which presented to reason nothing that it could not easily conceive, was a broad foundation for a popular and universal religion,—an advantage which Mohammed fully appreciated. With the Jews, who clung to their abrogated ceremonial, he maintained the authority of the Pentateuch, and the inspiration of the prophets from Moses to Malachi. With the Christians he admitted the Divine mission of Christ, and the truth of his Gospel; for he made the revelations both of the Old and the New Testament a basis for his own pretensions.* But as the Arabs were the more immediate objects of his imposture, he took more than ordinary pains to conciliate their affections. While lamenting the madness and folly of the idolatries in which they were plunged, he showed an extreme indulgence to their prejudices. Their popular traditions and ceremonies—such of them at least as favoured his own views—he retained, and even rendered more attractive, by adding the sanction of Heaven to customs already hallowed by immemorial usage.

But the most pleasing of all his doctrines, and the most captivating to the human heart, was the felicity promised in another world. The Mohammedan paradise is one of the richest and most seductive

* For the Author of Christianity the Mohammedans are taught to entertain a high reverence. "Verily," says the Koran 'Christ Jesus, the son of Mary, is the Apostle of God and his Word, which he conveyed into Mary: honourable in this world and in the world to come, and one of those who approach near to the presence of God."—(Chap. iii. iv.) Yet they consider him a mere mortal, and allege that a criminal or a phantom was substituted for him on the cross.—*Koran*, ib., and *Sale's Notes*. *Ma*rac[illegible], in *Alcoran*, tome ii. p. 113, 173; and *Prodrom.* part iii p. 163.

fictions of oriental imagination. The elements of its happiness consist not in pure and spiritual pleasures. These were too refined, and quite unsuited to the sensual habits of the Arab. The unlettered barbarian cannot comprehend the nature of abstract enjoyment, or how it can be felt without the agency of the bodily organs. To these carnal ideas Mohammed addressed his allurements, painted in the gayest colours that a luxurious fancy could invent. Gardens fairer than that of Eden, watered by a thousand streams, cooling fountains, and groves of unfading verdure, adorned these happy mansions. The desires of the blessed inhabitants were to be gratified with pearls and diamonds,—robes of silk,—palaces of marble,—rich wines,—golden dishes,—blooming girls, made of musk, with black eyes, of resplendent beauty and virgin purity. While these costly and exquisite indulgences were provided for the meanest believer, the most excruciating torments that imagination could suggest were denounced against all who refused to embrace the faith of Mohammed. Seven hells, differing in the degree of their pain, were to receive the damned; and the wretched sufferer might judge of his terrible doom when he was informed that the tenderest of these punishments was to eat burning victuals and to be shod with shoes of fire, the heat of which would cause his scull to boil like a caldron.

One other artifice was wanted to give effect to this plausible system,—the sanction of a Divine authority. A succession of prophets had already appeared in the world to instruct and reprove mankind, ever prone to wander from the truth; all of whom had their credentials attested by Heaven In this catalogue of inspired teachers, Mohammed determined to enrol himself. It was a bold but a necessary policy; and accordingly, next to the UNITY OF THE DEITY, stands the second fundamental article of the Mussulman faith,—that MOHAMMED IS

THE APOSTLE OF GOD. On these two pillars,—the one an eternal truth, the other an impious fiction,—the Eastern imposture has rested with unshaken stability for upwards of 1200 years.

Having at length matured his plans, and acquired a reputation for sanctity corresponding in some measure with the high and venerable office he was about to assume, he now resolved to make his pretensions to revelation no longer a secret. His fortieth year was the period chosen for announcing his mission to the world. He had retired, according to custom, to the grotto of Mount Hara, accompanied with some of his domestics. It was on the 25th of the month Ramadan,—the night styled in the Koran Al Kadr, or the Divine Decree,—that he received his instalment into the apostolic office in "a true and nocturnal vision." The archangel Gabriel, his confidant and oracle in all his celestial communications, descended in a brilliant form. He held in his hand a book brought from the seventh heaven. "Read!" exclaimed the angel. "I cannot," replied his awe-struck pupil. "Read," added the other, "in the name of God, the Creator, who hath formed man, and taught him the use of the pen, and lighted up his soul with a ray of knowledge!" The Prophet obeyed; and a voice immediately pronounced these words:—"O Mohammed! thou art the apostle of God, and I am Gabriel." This joyful inauguration into his ministry was received in silent wonder; the angel, having performed his part, ascended slowly and majestically until he disappeared in the clouds.

The conductor of the Israelites had produced the Pentateuch, and the Redeemer of mankind had taught the gospel. This "last of the prophets," too, must have his book; and now, for the first time, the Koran descends to earth. It was one of the most skilful of his artifices, and to which he mainly owed his success,—that instead of communicating this celestial volume entire, as the archangel brought it, it was doled out in morsels as suited his convenience. This sage manœuvre gave him a complete mastery over the oracles of Heaven; for he could make them speak according to circumstances. The Roman pontiff, who at this very time (A. D. 606) had begun to assert his claim to universal supremacy, might boast of the keys of Peter; but Mohammed held the keys of Providence, with which he could shut or unlock the gates of revelation at pleasure.

This pretended interview with the archangel rested solely on the suspicious authority of his own assertion. The first person to whom he related the tidings was Kadijah. The dutiful wife believed, or affected to believe, the sacred fable, with all its glorious accompaniments; and with a solemn oath she declared her conviction that he was the true apostle of his nation. "Among men," said the Prophet on this occasion, "many have been found perfect; but of women only four—Asia, the daughter of Pharaoh; Mary, the daughter of Amran; Kadijah, the wife, and Fatima, the daughter, of Mohammed;" where it will be observed that, with singular modesty, he includes the half of these female paragons in his own family. The second proselyte was his cousin, Ali, the son of Abu Taleb, then only eleven years of age, whom he had brought up in his own family with a fatherly tenderness. His slave, Zaid, was the third convert. Whatever might have been his scruples, they were overcome by the promise of liberty; and the grateful domestic recognised with joy the divinity of a master from whom he expected and obtained his freedom.

The next and most important of his conversions was that of Abdallah, surnamed Abu Beker, an opulent citizen of Mecca. He was a most zealous Mussulman; and, being a person of great authority among the Koreish, he prevailed on five of the principal men in the city—Othman, Abdalrahman, Saad, Zobeir and Telha—to join the standard of the Prophet. These six individuals were his chief associates—his main instruments in disseminating his religion—the partners of his victories—and some of them his successors on the throne. Three years were thus spent in devotional intrigue and the secret process of discipleship, during which the shades of mystery were allowed to conceal from the world his doctrines and his crafty designs.

But the time had now arrived when he could rely with confidence on the attachment of his new proselytes. Immediately the angel commanded him to make known his sacred vocation, and to exhort his friends and neighbours in particular to forsake their errors, if they hoped to escape the vengeance of an offended Deity. The obedient apostle accordingly directed Ali to prepare an entertainment—a lamb and a bowl of milk—to which forty guests of the race of Hashem were invited. After some interruption, Mohammed addressed the astonished assembly—"Friends, I this day offer you what no other person in all Arabia can offer,—the most valuable of gifts,—the treasures of this world and of that which is to come. God has enjoined me to call you to his service. Who among you will be my vizier, to share with me the burden and the toils of this important mission—to become my brother, my vicar, and ambassador?" This address was heard with silent surprise; and none seemed disposed to accept the proffered dignity. At length the impatient Ali made answer—"I, O Prophet, will be your vizier, and obey your commands! Whoever dares to oppose you, I will tear out his eyes, dash out his teeth, break his legs, and rip open his body!" On this burst of enthusiasm, Mohammed caught the youth in his arms with the liveliest demonstrations of affection. "Behold," said he, "my brother and vicegerent! Listen, and obey him." Shouts of contemptuous laughter followed this romantic installation. The whole company turned their sarcastic eyes on Abu Taleb, as if to inquire whether the rights and honours of a father were to be violated by rendering obedience to the authority of his own son.

Far from being silenced by this ridicule, or discouraged by the unfavourable reception of his first public attempt, the intrepid apostle laboured with indefatigable zeal, and marched onward with unshaken resolution to the final accomplishment of his designs. No reproaches or affronts could damp his ardour, for he bore them apparently without resentment; while every artifice was employed to subdue opposition. But it was to the force of his natural eloquence as a preacher, and the fertility of his genius, that he mainly trusted. On solemn festivals, and in the times of pilgrimage, he frequented the temple, and accosted the strangers of every province. Their imaginations and their passions were alternately excited by threats and promises. To the believer the carnal enjoyments of paradise were liberally offered; while, for the infidel, collars, chains, and torments unutterable were laid up in store.

The people trembled for their gods, which already seemed toppling from their pedestals. The Koreish, especially, dreaded the effects of his zeal. They beheld the worship which was their chief means of support threatened with extinction; and they resolved to crush in the birth this attempt to sap the foundation of their wealth and consequence. A deputation of the principal men of the tribe laid their fears and complaints before Abu Taleb: "Unless thou impose silence on thy nephew, and check his audacity, we shall take arms in defence of our gods. The ties of blood shall not restrain us from drawing the sword, and we shall see on which side victory will declare itself." Alarmed at these menaces, he seriously exhorted the Prophet to abandon his rash and impracticable schemes. "Spare thy remonstrances," said the daring fanatic; "though the idolaters should arm against me the sun and the moon, planting the one on my right hand and the other on my left, it would not divert me from my resolution." Meanwhile the Koreish, finding that neither threats nor entreaties could prevail, in a public assembly of their whole tribe, pronounced sentence of exile against all who had embraced the religion of Islam.

Five years of the Prophet's mission had elapsed; and his success may be estimated from the tradition that, of his disciples, who were now compelled to seek refuge in voluntary banishment, only eighty-three men and eighteen women, besides children, retired to Abyssinia.* While he remained exposed to this tempest of indignation at Mecca, a fortunate accident brought an important accession to his party in the conversion of two distinguished individuals—the brave Hamza, one of his uncles, and Omar, the second of the caliphs. The Koreish had secretly plotted his death; and there was only wanted an arm bold enough to strike the blow. The ferocious Omar agreed to be the assassin; but whether through the remonstrances of a friend, or, according to Abulfeda, by hearing a few sublime verses of the Koran read, instead of plunging his dagger in the breast of the apostle, the murderer was transformed into one of his most devoted proselytes.

For years the Koreish had beheld with jealousy the rising pre-eminence of the family of Hashem. Zeal for their national religion imbittered those political animosities, and served as a cloak to cover their malice. A solemn decree was passed in the name of the whole tribe, engaging themselves to

* On this subject the Abyssinian annals are silent; and the Mohammedan statement has been considered a fiction.—*Ludof.* in *Comment. ad Hist. Ethiop.* p. 284. *Maracci*, in *Prodrom.* p i. cap. 2. *Gagnier, not. ad Abulfed.* p. 24. *La Vie de Mah.* c. x.

renounce all communication with the Hashemites neither to buy nor sell with them, to marry nor give in marriage; but to pursue them with implacable enmity until they should deliver up this dangerous innovator to the resentment of the nation, and the justice of the gods whose worship he had deserted. The deed was written on parchment, and suspended on the wall of the Kaaba, that all eyes might read it.

Having no security in the city, the persecuted faction withdrew to a stronghold in the neighbourhood. Here they remained three years in a state of siege; the only intervals of their captivity being the sacred months, when hostilities were prohibited. During the ceremonies of the pilgrimage, the two factions regularly met, and frequently came to blows. The orations of Mohammed in the temple were often drowned amid the clashing of swords and the exhortations of the idolaters in behalf of their ancient divinities.

Hitherto the credit of Abu Taleb had been the main asylum of the apostle and his followers, and was perhaps the true cause of rescinding the prohibitory edict, after it had subsisted five years. Death deprived him of that support; and within a month this domestic calamity was followed by another,—the loss of Kadijah in her 65th year. The Prophet was inconsolable; for he had always regarded her with ardent and undivided affection. During the five-and-twenty years of their marriage his fidelity was irreproachable; and the rights or feelings of the wife were never insulted by the society of a rival. His tears and praises spoke his sorrow long after she was in the grave; and his excessive encomiums wounded the pride of her successor, the youthful Ayesha. "Was she not old," said the petulant and blooming daughter of Abu Beker, "and has not God given you a younger and a better in her place?"—"No, truly," replied the grateful apostle, "there never can be a better; she believed in me when men despised me. She was generous, and gave me all she possessed, when the world hated and persecuted me." Misfortunes so distressing and prejudicial to his interests made the Mohammedans commemorate this as the Year of Mourning.

A valuable accession was about the same time received to his flock in a small party of the tribes of Khazraj and Aus, who dwelt at Medina, and had come to Mecca on the usual pilgrimage. The secret motive of their conversion was a hope that their new master was the long-expected Messiah, and would deliver their allies, the Jews, as he had promised, from the vassalage to which they had been so long subjected. On their return these deluded proselytes became enthusiastic in disseminating so welcome a creed among their fellow-citizens.

Historians, or rather the lovers of the marvellous, have signalized this period of Mohammed's life with two remarkable events, the absurdity of which might have consigned them to oblivion, had not the gravest of the Moslem doctors maintained their reality. Religion, whether true or false, has usually appealed to the confirmation of miracles. These credentials the impostor himself admitted to be authentic. According to his own doctrine, therefore, the unbelieving Arabs might demand, and they did repeatedly urge him to produce, similar evidence of his mission. Sensible of his weakness, he evaded the force of their objections—appealing to the inimitable composition of the Koran as the greatest of all miracles, and protecting himself by the obscure boast of vision and prophecy.

His votaries, however, were neither so modest nor so ingenious. Of his miraculous gifts they were more confident than he was himself; and much learning has been expended, and innumerable volumes written, to convince the world that his miracles were more numerous than those of all the inspired teachers who had gone before him.

The first of these signal performances was the miracle of the Splitting; alluding to his cleaving the orb of the moon in twain. The Koreish, wishing to confound him before the eyes of his fellow-citizens, had challenged him to verify his claims by bringing that luminary from heaven in presence of the whole assembly. Mohammed accepted the proposal with confidence. At his command the sky was darkened at noon; when the obedient planet, though but five days old, appeared full-orbed, leaped from the firmament, and, bounding through the air, alighted on the summit of the Kaaba, which it encircled by seven distinct revolutions. Turning to the Prophet, it did him reverence, addressed him in very elegant Arabic, and pronounced a discourse in his praise, concluding with the formula of the Moslem creed. These salutations finished, it entered the right sleeve of his mantle, and made its exit by the left. Then descending from the collar of his robe to the fringe, it mounted intò the air, separating into two halves. In this manner it resumed its station in the sky, the parts gradually uniting in one round and luminous orb, as before. Such is the substance of a ridiculous fiction invented by the biographers of Mohammed, who have coloured it with more extravagance and minuteness of detail than we have ventured to narrate.

The next legendary adventure of the Prophet is yet more extraordinary—the *Mesra*, or famous nocturnal journey to heaven; of which the Eastern writers, in the wild delirium of their fancy, have given the most laboured and grotesque descriptions. With sublime touches of imagination, that would have done honour to the muse of Milton or Dante, they have mixed a legion of idle phantoms and puerile wonders too shocking and extravagant even for the credulity of childhood.

On the night of this celestial excursion, calm but exceedingly dark, Mohammed represents himself as asleep between the hills of Safa and Meroua, when Gabriel approached and awoke him. Having apprized the prophet of his intended voyage, he presented him with the animal called Borak, a sort of nondescript, larger than an ass but smaller than a mule, with a human face and the body of a horse. His colour was milk-white; the hair of his neck of fine pearls; his ears emeralds, and his eyes two sparkling hyacinths. His whole body, wings, and tail, bristled with the finest jewelry.

In the twinkling of an eye they cleared the hills of Mecca, and were on the top of Sinai, where prayers were said, and where the print of the beast's hoof is still shown. In the same manner they performed their devotions at Jerusalem, where Mohammed received the salutations of the ancient prophets, and met with divers other adventures. Leaving Borak fastened to a ring at the gate of the temple, the travellers ascended by a ladder of light, through an immense expanse of air, till they reached the first heaven, distant a journey of 500 years from the earth. It was composed of a subtle vapour, with a roof of fine silver, from which hung the stars by chains of massive gold. They entered by a prodigious gate, which on the name of Mohammed being announced, was opened by the porter. The first person with whom he exchanged salutations was Adam,* who appeared in the form of a decrepit old man, and hailed him as the greatest and best of his posterity. The whole firmament swarmed with angels all busy in their several occupations, some watering the clouds, others chanting hymns. They appeared in all manner of shapes,—men, beasts, and birds, for each assumed the likeness of those

* Mohammedan authors differ in the location of the patrirchs Few will dispute, and we have not thought it import- ontest points of fabulous precedency.

terrestrial creatures intrusted to their spiritual guardianship. The most conspicuous of these was the angel or representative of the cocks, white as snow, and of such gigantic stature that his head touched the second heaven (a distance of 500 years travel); or, as others affirm, reached through all the seven heavens. He assisted in the matin songs of the angelic choirs, and gave the signal for all his species to crow, whether material or immaterial.

The second heaven was of pure gold, and contained twice as many angels as the first. Here Mohammed was saluted by Noah, who commended himself to his prayers; but he was not permitted to take further notice of the various marvels he saw. The third heaven was made of precious stones, and more populous than the second. Here the travellers were greeted by David and Solomon, and saw a huge angel called the Faithful of God, who had 100,000 others under his command. In the fourth heaven, which was of emerald, they received the felicitations of Enoch and Joseph. Here they beheld an angel of a very stern and terrible aspect; the distance between whose eyes was equal to 70,000 days' journey according to the rate of Arabian travelling; and such was his capacity, that he could have swallowed the seven heavens and seven earths as easily as a pea. Before him was a large table on which he was continually writing; inserting the names of all that were born, computing the days of their lives, and blotting them out from his register the moment their allotted portion of years expired. It was Azrael, the angel of death, whose emissaries traverse the earth perpetually, keeping watch over the issues of human life. No smile ever lighted up his dismal visage; his business being to weep and make lamentations for the sins of men.*

* Prideaux condescends (and so does Maracci) to grapple in serious combat with this phantom. As the distance between a

Into the fifth heaven, which was composed of adamant, they were admitted by a gate of pure silver, inscribed with the Mohammedan creed. Aaron gratulated them on their arrival. This sphere was the great storehouse of God's wrath;—a black and horrid pit, vomiting forth a thick smoke, the stench of which was insupportable. The presiding angel of this infernal treasury was hideously deformed, his withering look being enough to blast the material universe. His eyes were of rolling flame, his face like copperas, disfigured with wens and excrescences; and around him lay darts and chains of fire, the terrible instruments of divine vengeance, which were kept in constant preparation for rebellious sinners,—especially for the unbelieving Arabs. Quitting these dreary mansions they advanced to the sixth heaven, which was of carbuncle. At some distance they perceived an aged man, with shaggy hair, clothed in a woollen garment, and leaning on a staff. It was Moses who saluted his brother prophet; but immediately burst into tears at the thought that this "Arabian boy" would be instrumental in bringing more of the race of Ishmael into paradise than he and all the prophets had done of the Jewish nation. Here they met with another prodigy in pneumatology,—an angel, one half of whose body was snow and the other fire; yet these discordan elements were neither melted nor extinguished.

But the most marvellous of all created beings was the tutelar angel of the seventh heaven. He hac

man's eyes is in proportion to his height as 1 to 72, he calculates that this angel must have been four times the length of all the seven heavens, and therefore could not stand in one of them. "Here," says he, "Mahomet was out in his mathematics." But a captious Mussulman might argue with the dean that the angel was not intended to *stand*, but to *sit;* for he told Mohammed he had not permission to quit his desk from the creation of man till the final judgment.—*Prid. Life*, p. 61. See also *Buxtorf's Synag. Jud.* cap 50, and *Purchas' Pilgrims*, lib. ii. cap. 20

70,000 heads, each head 70,000 faces, each face as many mouths, each mouth as many tongues, and each tongue spoke seventy thousand different languages, all of which were employed incessantly in praise and adoration. This last and highest of the celestial spheres was made of divine light. Here was the abode of Abraham; and, according to some of Jesus Christ, who is alleged to have treated Mohammed with the same respect as the other prophets.*

Having penetrated to the lotus-tree (Al Sedra) which is the utmost limit of created knowledge, the boundary of these delicious regions, beyond which no angel dares to pass, Gabriel took leave of his fellow-traveller, commending him to the protection of superior spirits during the remainder of his journey. Continuing his march through ranks of glorified cherubim, and crossing two seas, one of light and one of darkness, the solitary prophet passed the 70,000 veils of separation, each being a journey of 500 years in thickness, and the same in distance between them. They were composed, some of darkness, others of fire, snow, water, ether, and chaos. Finally, he pierced the veils of beauty, of perfection, of omnipotence, of singularity, of immensity, and of unity. When the last of these was raised, 70,000 spirits were seen prostrate before the throne, which was surrounded by a light of the most dazzling brightness. A voice commanded him to draw near; on which he advanced till within two cubits, or bows' length, of the Divine presence. As a mark of his favour, the Almighty, we are informed, laid his hand on the prophet's shoulder, when a feeling of intense cold thrilled to the mar

* Prideaux thinks, in this latter instance, the prophet altered his style of salutation, acknowledging the superiority of the Messiah,—a supposition at variance with the doctrine taught in the Koran

row over his whole frame; but was immediately succeeded by a sensation of inexpressible sweetness. This was followed, as he pretended, by a long and familiar intercourse with the Supreme Being, who revealed to him many hidden mysteries, instructed him in the knowledge of his law, and conferred on him several extraordinary privileges. The last of his instructions was the command of fifty daily prayers, afterward reduced by the advice of Moses to five, enjoined on all Mussulmans.

Bidding adieu to these glorious regions, Mohammed rejoined his conductor Gabriel, whom he found by the lotus-tree. The travellers now bent their course towards the earth, receiving everywhere, as they passed, the compliments and benedictions of angels, who flocked in crowds to salute them. At Jerusalem they found Borak in the exact position they had left him; and in less than a second they arrived at Mecca,—the slumbering inhabitants being quite unconscious of the transactions of that marvellous expedition; for the whole journey, the labour of so many thousand years, was performed in the tenth part of a night. Such is the celebrated *History of the Ascension*, as Abu Horaira calls it, whose minute and circumstantial account we have abridged from Gagnier.

A controversy arose, and continued long to divide the Mohammedan world, whether the nocturnal voyage was a real and corporeal journey, or merely a vision. Ayesha, his wife, maintained (or tradition in her name) that the prophet never left her bed, and that his spirit, and not his body, travelled. Some compromised the miracle, by admitting a real translation of the body from Mecca to Jerusalem, but regarding the ascension itself as a dream. Others strenuously maintained the corporeity of the whole voyage from beginning to end; declaring that to deny this was a damnable error, and as much an act of infidelity as to reject the Koran. The

Turks celebrate the 20th night of Rajeb by a gran festival, in commemoration of this event.*

While Mecca was filled with disputes on the nocturnal voyage, and ridiculing its author as dreamer and a visionary, the streets of Medina resounded with his praises. The zealous converts there, twelve of whom had been vested by Mohammed with apostolical authority, had animated others with their own enthusiasm, and considerably multiplied the number of proselytes.

Mosaab their chief repaired to Mecca, at the time of the pilgrimage, with seventy-three men and two women, all eager to do obeisance at the feet of their master, and proffer him their assistance. At a private conference Al Abbas explained to them the persecutions his nephew had suffered on account of his opinions; the necessity of abandoning his native place to seek protection elsewhere; and the favourable asylum which seemed to present itself in their generous proposal. "And what will be our recompense should we fall in the quarrel?" "Paradise!" This single word fixed their determination. They expressed their resolution to defend, and never to betray him; and took an oath to commit no vice, and to protect him with life and fortune, "as they would their wives and children, against all nations, black and red, who should dare to oppose the faith and the apostle of the Koran." To this solemn pledge of mutual fidelity, called the Women's Oath, may be traced the first vital spark of the Saracen empire.

The altered fortunes of the impostor changed the course of his policy, and his ambition seemed to drop its veil in proportion as the means of support increased. Hitherto his pretended revelations had spoken nothing but the language of peace and for·

* Vide Mod. Univ. Hist. vol. i. p. 80, and authorities quoted there. Koran, chap. xvii. liii

bearance. But no sooner did a concurrence of favourable circumstances draw over a sufficient party to his views, and open a friendly retreat in the heart of a warlike city, than he threw off the mask, and resolved to substitute a mode of conversion less tedious and uncertain in its operation than the gentler arts of argument and persuasion. The permission of Heaven to take up defensive arms was changed into the stern command to make war. Chapter after chapter descended, to encourage the faithful in the work of extirpation; for Gabriel, who had withheld these injunctions so long as discretion was the better part of valour, was now ready to quicken the sacred process of excision by new revelations. "Make war against unbelievers—strike off their heads, and strike off the ends of their fingers. This shall they suffer because they have resisted God and his apostle." (Chap. viii. and xlvii.)

Of the secret confederacy with the Medinians the Koreish were soon apprized; and they foresaw the possibility of immediate invasion from the two most warlike tribes in their vicinity. An extraordinary council was assembled, headed by Abu Sofian, who had usurped the sovereignty of Mecca at the death of Abu Taleb, and thus excluded the family of Hesham from the throne. The proposition of Abu Jehel was carried, that Mohammed should be put to death by assassination. A man from each tribe was to be selected, who with their poniards were to execute the sentence while he lay asleep; and by imbruing so many hands in the guilty deed, they hoped to overpower any attempts on the part of his kinsmen or followers to avenge his death. The devoted apostle had timely information (by the agency of Gabriel, or rather the aid of a human spy) of this conspiracy, though the following night was to have witnessed its execution. He communicated the secret to the generous Ali; and having instructed him to wrap himself in his green mantle and lie down in his place, he took his departure; eluding by this stratagem the vigilance of the assassins, who had already planted a guard at the door. Favoured by the darkness of the night, Mohammed reached the house of Abu Beker in safety. Without delay these two fugitives left the city on foot; and, to lull suspicion, repaired with a hired guide to the cave of Thor, a hill three miles from Mecca, where they lay concealed for three days, receiving in the twilight of each evening, from Abdallah and Asama, the son and daughter of Abu Beker, a secret supply of intelligence and food.

Meantime the assassins, perceiving through a crevice their supposed victim, and waiting in anxious silence the approach of slumber, remained on the watch till morning, when they were undeceived by the appearance of Ali, whom they allowed to escape unmolested. The intrepidity of the heroic youth commanded the respect of the Koreish, and was made the subject of exalted panegyric by the Moslem historians, who have held up his fidelity in exposing his own life to save that of his benefactor as an example which angels were recommended to imitate. Two of these spiritual messengers, we are told, were stationed near his bed, the one at the head, the other at the foot; but their presence seemed to have failed in quieting his agitation; for, in some verses still extant, he has expressed with considerable pathos the conflicting emotions of hope and fear, of tenderness for his friend and confidence in religion, which filled his bosom on that occasion with perplexing suspense.

Stung with rage and disappointment, the Koreish sent spies and armed parties to explore every haunt in the neighbourhood of the city; and offered a reward of a hundred camels to any man that should take the adventurer alive or dead. They arrived at the cavern, whose terrified inmates overheard their conversation. "We are only two," said the trem bling Abu Beker, who had shed many bitter tears at the desperate fortunes of his master. "There is a third," replied the undaunted prophet,—"It is God himself." A cherished tradition of the Arabs has invented a providential deceit, which saved the fugitives,—a pigeon's nest with two eggs, and a spider's web drawn completely across the mouth of the passage. Convinced from these appearances that the place was solitary and inviolate, the pursuers desisted from all further examination. The virtues of this miracle, Mohammed used to say, were better than a coat of double armour in defending him from the swords of his enemies.*

The greatest impatience was manifested at Medina for the coming of the prophet. At the suburbs he was met by 500 of the inhabitants, who received him with every possible demonstration of joy. Here he was again joined by the faithful Ali, within three days after his arrival. The day of his entering the town, which now changed its name of Yatreb for that of Medina (or Medinat el Nebbi, the city of the Prophet), is generally admitted to have been the 16th of Rebiah I., he having left Mecca on the first of that month.†

* Al Damiri, in his History of Animals, assures us, that in memory of this event, the pigeon was held sacred by the Mussulmans; and that, for the same reason, Mohammed forbade the killing of spiders. Some allege, that the fable of the pigeon whispering revelations into his ear, to persuade his followers, as the Christians relate, that he was divinely inspired, took its origin from this circumstance.—*Pococke, Spec.* p. 186. *Reland. de Relig. Moham.* p. 359. *Gagnier, La Vie de Mahom.* p. 290. *Grotius, de Veritat. Relig. Christ.*

† The departure of the prophet has fixed the memorable epoch of the Hejira, or Flight, the era by which the Mohammedan nations still compute their lunar years. Like that of the Christians, it was not introduced until some time after the death of its founder. Its appointment belongs to the Caliph Omar, who, being appealed to in a controversy between a debtor and his creditor, the former alleging that the month mentioned in the bill did not belong to the current year, but to the following, and con

The religion of the Koran, after struggling with thirteen years of misfortune, might have withered in the bud, had it not struck firm root in the loyalty and devotion of the Medinian converts. The first care of the apostle was to erect a place of worship, where he might publicly discharge the sacred functions of his office. A small parcel of desolate ground was purchased by Abu Beker, and on this chosen spot were founded a mosque and a house for the prophet. It was the patrimony of two orphans; and the enemies of Mohammed have gravely but falsely accused him of despoiling the helpless children of their inheritance.*

To hasten the completion of this venerable structure, Mohammed laboured with his own hands; it was merely a rude chapel with mud walls, on which was placed a roof of palm-leaves, supported by the trunks of date-trees for pillars. Near it was built a house for his youthful bride, Ayesha, then only in her ninth year,—such is the premature ripeness of Eastern climes. Already the Prophet had divided his affections between her and Sawda, one of his

sequently that the money demanded was not then due, ordained, that to remedy all such inconveniences, their computations in future should begin with the flight of the apostle from Mecca. This new epoch, however, made no alteration in the ancient form of the Arabian year, which commenced, as before, on the first of the month Moharram, fifty-nine days earlier than the departure of the prophet. But in order to simplify their calculations, the Hejira was made to precede the real event by fifty-nine days, and is generally supposed to correspond with Friday, the 16th of July, in the year of our chronology 622.

* Prideaux's story of the robbery of the poor orphans, sons of a *carpenter*, and the injustice of the impostor, is shown to be erroneous. Al Najjar, which he translates carpenter, was the name of a rich and noble tribe. Gagnier adduces the authority of Jannabi and Bokhari, that the ground was offered him in a present, which he refused; and that Abu Beker paid the money out of his own pocket. "Ils lui disent, 'O Apôtre de Dieu, nous vous le cedons en pur don.' Le Prophète voulut absolument l'acheter, et Abu Becre le paya de ses propres deniers."—*La Vie de Mah.* p. 302.

earliest disciples; and shortly after he espoused Haphsa, the widowed daughter of Omar; thus confirming his interests by forming matrimonial connexions with three of the principal men of his party. The etiquette of a separate habitation was a mark of attention which he paid to all his wives; and in a brief space, the new temple saw its precincts adorned with nine of these conjugal mansions.

The next and most essential object of Mohammed was to amalgamate the jarring elements of his congregation. The Medinian proselytes had received the honourable title of Ansars, or Helpers, and the exiles of Mecca took the name of Mohajerin,—Refugees, or companions of his flight. To eradicate the seeds of jealousy which this distinction might create, both parties were bound by a fraternal league, not only to live in peace and concord, but to love and cherish each other with the tenderness of brothers. As an additional tie, he joined them in pairs, each refugee being coupled with an auxiliary companion. This expedient was completely successful. The holy brotherhood respected their obligations both in peace and war, and during the life of their master vied with each other in a generous rivalry of loyalty and valour. Once only in an accidental quarrel was the voice of discord known to interrupt their affectionate union; but such was the stern spirit of their fidelity, that the believing son offered to lay the head of his idolatrous and offending father at the apostle's feet.

The second year of the Hejira was ushered in with the institution of certain external rites of the Mohammedan worship. To gratify the Jews, the *kebla*, or point to which they turned their faces in prayer, was fixed in the direction of Jerusalem; but in trying to ingratiate himself with the Ansarian party, the Prophet greatly displeased the Arabs, whom nothing could wean from their respect for the Kaaba. Again the omnipotence of revelation was called in;

and henceforth all true Mussulmans were commanded, from whatever quarter they might come, to turn their faces to the *Haram*, or Holy Temple of Mecca. In order to silence heretics and revilers, whose objections had so teased and perplexed him at his outset, Mohammed resolved to interdict in future the presumption of doubting his mission, or disputing about his religion. Death was the award in the Koran for all who should dare to contradict or oppose any of the doctrines he taught. Fighting and not controversy was now to be the only legitimate mode of propagating the true faith; and its opponents, of whatever creed, must either believe at the point of the lance, or redeem their lives by submitting to pay an annual tax for their infidelity.

The enthusiasm of the Arabs was thus doubly inflamed, by the hope of plunder and the promise of a sensual paradise. The decrees of an absolute fate, which would extinguish both industry and valour if men were left to the influence of a merely speculative belief, were dexterously turned into instruments for inspiring the disciples of the Koran with the most exalted and reckless courage. The companions of the Prophet advanced to battle without fear. As nothing was left to chance, there was no room for danger or dismay. The same inevitable destiny that might have ordained them to perish in their beds, would not overtake them a moment sooner on the field of death, or render their persons more insecure amid the arrows of the enemy. The lot of all was determined by a fixed and resistless predestination; with this difference, that while the man of peace departed obscure and inglorious, the fallen warrior had before his eyes the crown of martyrdom and the joys of paradise. "The sword," exclaimed the military apostle, "is the key of heaven and of hell! A drop of blood shed in the cause of God, a night spent in arms, is of more avail than two months' fasting or prayer. Whoever falls in battle his sins are forgiven; at the day of judgment his wounds shall be resplendent as vermilion, and odorous as musk; and the loss of his limbs shall be supplied by the wings of angels and cherubim." The valiant martyrs of the faith were allowed to anticipate the voluptuous enjoyments of another world, by the license of embracing the female captives as their wives or concubines. The interval of the four sacred months, which had hitherto suspended the fury of the most hostile tribes, was disregarded, that no impediment might retard the victorious Moslems in their mighty career of pillage and proselytism. The distribution of the spoil was regulated by the authority of revelation. The whole plunder of the forage or the battle-field was to be collected in one common mass. A fifth part the Prophet reserved to himself for charitable and pious uses; the remainder was to be divided among the soldiers, including those who guarded the camp as well as those who had been actually engaged. The portion of the slain devolved to their widows and orphans; and to encourage the increase of cavalry, each horseman was allotted a double share.

The hostile principles inculcated by the Koran did not long remain a dead or speculative precept. In the twelfth month after his settlement at Medina the despised and persecuted outcast of Mecca proclaimed a Holy War against the Koreish.* Various ambuscades were stationed to annoy their commerce, by seizing the caravans as they winded through the narrow defiles of the mountains. Parties of three or fourscore horsemen continued to reconnoitre month after month without gaining any important advantage. But the failure of these preliminary attempts was speedily redeemed by Mohammed himself on the plain of Bedr, one of the

* Jannabi, cited by Gagnier, seems to hint that the Koreish were the first aggressors.—*La Vie de Mah.* tome ii. lib. iii chap. 2.

usual watering stations, about forty miles from Mecca. Spies had brought him intelligence that a caravan of the idolaters, consisting of about 1000 camels richly laden with grain, fruit, and other costly merchandise, was on its return from Syria, guarded with an escort of only thirty or forty men, commanded by Abu Sofian in person. Persuaded that this valuable and apparently easy prey was within his grasp, he resolved to advance at the head of a small detachment of troops to intercept it. This sacred band of warriors did not exceed 313 men, of whom seventy-seven were Mohajerins, and the rest chiefly Ansars. So poorly were they accommodated in regard to cavalry, that they could muster only two horses and seventy camels, which they mounted by turns. The plan of their future operations being decided, the leader of the faithful advanced, and pitched his tents at a short distance from the enemy. A slight intrenchment was thrown up to cover the flank of his troops; and for the safety of his own person, a temporary structure of wood, overshadowed with green boughs, was erected, with a fleet camel standing ready harnessed, that in case of defeat he might avoid the chance of being taken prisoner; for, however assured the Prophet might be of Divine assistance, he had too much sagacity to despise the use of human means. Burning with zeal and mutual hatred, the troops on both sides rushed furiously to the charge. The idolaters were three to one; but the superiority of numbers was overbalanced by the reckless intrepidity of fanaticism.

While the Moslems nobly sustained the assault of their adversaries, their commander fervently addressed Heaven in their behalf. Seated with Abu Beker in his wooden sanctuary, with his eye fixed on the field of battle: "Courage, my children, and fight like men!—close your ranks, discharge your arrows, and the day is your own. O God! execute what thou hast promised;" alluding to the celestial reinforcement which he had demanded of Gabriel. In this manner he continued in great perturbation to wrestle with Providence till the mantle fell from his shoulders. Then starting as from a trance,—" Triumph, Abu Beker! triumph! behold the squadrons of Heaven flying to our aid!" It is not improbable he had observed his little army beginning to waver or give way, and adopted this pretext for rekindling their enthusiasm. At that decisive moment he mounted his horse, placed himself at their head, and in a few verses of the Koran announced the arrival of their celestial auxiliaries. The Mussulmans were inflamed with renewed ardour, and imagined that the heavenly militia were to fight heir battles. The Koreish were dismayed and fled, leaving seventy of their warriors dead on the field, and seventy prisoners in the hands of the Faithful.

The glory of this first victory of the Moslems, the Koran has more than once piously attributed to the effect of Divine assistance. Their historians relate that the angelic chivalry, with Gabriel at their head, did frightful execution with their invisible swords on the terrified idolaters; though we cannot help thinking a smaller number than 3000 (others say 9000) might have sufficed for the destruction of threescore and ten of the Koreish. Such stories must to us appear idle and ridiculous; but they were the fuel with which Mohammed inflamed the martial enthusiasm of his army. It was by fostering the idea of God being their protector that he rendered them invincible; and such was the empire he had obtained over their imaginations, that he found it his interest to attribute to miracles the remarkable success which arose from the blind fanaticism wherewith he had himself inspired them.

The capture of their last caravan had determined the Meccans, in their next journey to Syria, to take the eastern route, through the desert, and along the borders of Irak. But it was in vain that their camels explored a new road. The banks of the Euphrates were not more secure than the shores of the Red Sea. The valiant Zaid, with a body of 500 horse, had orders to intercept and seize this wealthy prize; and so gallantly did he execute his commission, that he overtook the enemy at Al Karda, in the desert of Nejed; where, after defeating the escort, he made himself master of the entire caravan. The value of the plunder may be conjectured from the fact, that of the money alone the apostle's fifth amounted to 20,000 or 25,000 drachms.

These repeated losses filled the Koreish with shame and rage. The effusions of the muse were employed to stir up the passions of the indignant citizens. Caab, a Jew and an inveterate enemy of Mohammed, inveighed bitterly against him in satirical verses—an imprudence that cost him his life, as he was soon after assassinated by an emissary of the Prophet. The moment of excitement was favourable; and Abu Sofian speedily collected a body of 3000 men, of whom 200 were cavalry, and 700 were armed with cuirasses or coats of mail. Their march was attended by his wife Henda and fifteen other matrons of the Koreish, carrying timbrels and acting the part of drummers. To animate the courage of the troops they sung the elegies of Caab, lamenting the disasters of Bedr, and exhorting their warriors to fight valiantly. They proceeded without meeting the least resistance, and encamped at a village within six miles of Medina.

Mohammed was apprized of their approach; but, as he could oppose them only with a very inferior force, he thought it more advisable to await their attack within the walls of the city, than hazard an engagement in the open plain. Most of the officers, however, espoused a contrary opinion, and demanded to be led to battle. After morning prayers, the army of the Believers, amounting to about 1000 men, of whom 200 were cuirassiers, left the capital, and encamped on the base of Mount Ohud, at the distance of six miles to the northward. The dauntless apostle, though deserted by fifty of his followers, disposed of his troops to the best advantage. Having no cavalry,—for, except the horse on which he rode, there was not one in the whole army,—he posted fifty archers in the rear; fearing he might be surrounded by the enemy, who were at least three times his number. The cuirassiers he placed in the centre; and having made these dispositions, he ordered the whole line to await calmly the signal of attack. The archers, on whom chiefly the fortune of the day depended, he strictly commanded not to quit their position, even should the front give way. The Koreish drew up in form of a crescent; the center commanded by Abu Sofian, and the right by the famous Khaled, the bravest and most successful of the Arabian generals. The rearguard, or body of reserve, was under the surveillance of the heroic Henda and her matrons, who cheered the standard-bearers as they passed.

Both armies stood facing each other. At the word of attack, the Moslems fell upon the idolaters with a fury that nothing could withstand. The weight of the charge broke their centre, drove them down the hill, and might have secured to the Believers a bloodless victory, had they attended to the orders of their able commander. But the whole advantage was lost by the impatient rapacity of the archers. Elated at this first instance of success, and hurried away with the avidity of plunder, they abandoned the important station that had been assigned them. Their dispersion left the Mussulman army entirely unsupported, and destitute of its chief defence;—a circumstance which did not escape the practised eye of the intrepid Khaled. Seizing the favourable moment, he made a rapid wheel with his cavalry, and attacked the Moslems, flank and rear, with such bravery that they were soon thrown into a state of complete disorder, and exposed to the carnage of a ruthless and vindictive foe. To terrify them still more, he raised the cry that Mohammed was slain. Courage and presence of mind forsook the believers. The rout became general; nor could the voice or example of the Prophet, who fought with desperate valour, rally for a moment his broken and discomfited troops. Surrounded with a few of his bravest soldiers, he contested the victory with a heroism worthy of a better cause. Firm and cool, he exposed his person freely wherever the danger appeared greatest. He was assailed by showers of stones, arrows, and javelins; and saw many of his gallant officers wounded by his side. When Mosaab fell dead at his feet, he seized the standard and planted it in the hand of Ali. The ferocious infidels had penetrated to the spot where he stood, encouraging his generous followers, who had formed a guard or rampart around him. In the tumult of the affray, he was struck from his horse, wounded and bleeding to the ground; his face was dangerously pierced by ten javelins, whose iron heads stuck in the wound; two of his teeth were beaten out; his lip cleft to the bone; and his life itself must have fallen a sacrifice had not Telha, nephew of Abu Beker, received a blow levelled at his master, which shattered his arm so as to deprive him of its use ever after. Yet, in the midst of confusion and dismay, he calmly reproached the impious Otba for staining the visage of a prophet with blood, and blessed the friendly hand that stanched his wounds, and conveyed him to a place of safety.

Finding the rumour of his death a false alarm, Othman and a chosen body of adherents returned to the charge, and with the most determined valour succeeded in rescuing the apostle from his furious assailants, and bore him to a village in the neighbourhood, where he obtained the necessary refreshments of water and repose. The day was totally lost. The Moslems numbered seventy martyrs. At their head was the brave Hamza, who was secretly stabbed to the heart with a lance by a slave, in the commencement of the action, while fighting among the foremost. Abu Beker, Omar, and Othman were wounded. The assurance of paradise was the reward of the fallen; while seventy-two prayers obtained for Hamza a place among the inhabitants of the seventh heaven, with the glorious title of the Lion of God.

The infidels remained masters of the field; but the orderly retreat of the Moslems deterred them from attempting pursuit, or taking advantage of their success. They stripped the slain, committing on their senseless trunks the most revolting excesses of vengeance. Their noses and ears were cut off, and worn in triumph by the victors, as necklaces, bracelets, and belts. Henda, recovered from her panic, with a barbarity rare even among savages, tore out the entrails of Hamza, gnawed his liver with her teeth, and swallowed part of the bloody morsel. Abu Sofian cut slices off his cheeks, and hoisted them on the end of a spear; shouting praises to Hobal, their popular deity, and his victorious religion. This brutal exultation might satiate their fury, but it lost them the best fruits of the action. Instead of glutting their revenge by a useless cruelty, the Koreish might have followed up their success by marching on Medina, then in a state of weakness and mutiny, owing to a quarrel with the Jews. The pillage of that capital, the strongest motive in Arabian warfare, would have supplied them with fresh courage; a few hours might have put an end to the rising empire of the apostle and the Koran, and again restored to the Kaaba the allegiance of its revolted worshippers.

This disaster threatened to annihilate the Prophet's

reputation. With his usual confidence he had predicted the entire overthrow of the idolaters; and presented one of his officers with a sword, on the blade of which "certain victory" was engraven. The Moslems, unaccustomed to reverses, were greatly chagrined. Some murmured at the loss of their friends and relations; others expressed doubts as to his pretensions to the Divine favour; since, had he been a true apostle, Heaven, they said, would not have suffered the infidels to triumph over him in battle. To these objections Mohammed had a ready answer. The clamours of those who were not altogether satisfied with the sublime doctrines of eternal fate and the felicities of martyrdom, he put to silence by throwing the whole blame and disgrace of the loss on their own sins.

Since the treaty of alliance between the Jews and Moslems, at their settlement in Medina, they had lived in peace and harmony, enjoying mutual liberty of conscience, and all the privileges of free citizens. An insult to an Arabian milkmaid interrupted this cordiality, and occasioned a war of extermination against the people of the synagogue. The tribe of Kainoka were driven into exile, and all their property confiscated. The Nadhirites possessed a strong fortress three miles from Medina. To this Mohammed laid siege, and for six days it maintained an obstinate defence; but seeing no prospect of assistance, while their palm-groves were laid in ashes, they agreed to capitulate on condition of marching out with their lives, and as much of their moveables as a single camel could export. The remainder of the spoil fell into the hands of the besiegers; and, contrary to express law, instead of a fifth, Mohammed appropriated the whole booty to himself, to be distributed at pleasure. This stratagem was an expedient to recompense the devoted fidelity of the Refugees, now become his favourite disciples. The policy was not without danger; but the fifty-ninth sura (the Emigration) descended, expressly to ratify this monopoly, on the ground that neither horses nor camels were used in this expedition. Thus easily was Heaven made to contradict itself. One divine oracle superseded the obligations of another, and cancelled the pretensions of the whole army, in order that their crafty general might discharge his debts of private gratitude.

The fifth year of the Hejira beheld the territory of Medina violated by an allied army of Jews and idolaters, and the city of the Prophet menaced with utter destruction. For this hostile movement the vigilant apostle was not unprepared; and by the advice of Salman the Persian, he caused, for the protection of the city, a deep ditch or intrenchment to be dug round it. The hearts of the believers quaked to behold their suburbs covered with tents, and bristling with a forest of moving spears; but their general, concealing his own apprehensions, loudly reproached them with their want of faith. After twenty days of ineffective blockade, and finding their prospects of success entirely frustrated by divisions in their own camp, the confederate forces broke up the siege, and prepared to return home. These dissensions, it is generally believed were fomented by the emissaries of Mohammed, who had contrived to corrupt their leading men. The news of their precipitate departure was welcome intelligence to the Moslems, who commemorated this expedition as the War of the Ditch, or of the Nations, in allusion to the different tribes of which the allied army was composed.

Relieved from their formidable assailants, the faithful expected to rest from their fatigues, and enjoy the blessings of peace in the bosom of their families. The intention of their leader was very different. On the same day, and without laying aside his armour, Mohammed ordered his troops to march against the Koraidites, a Jewish tribe who had joined the confederates. The soldiers murmured: but it would have been impiety to disobey; for Gabriel is made to remonstrate with him for suffering his people to lay down their arms before the angels had laid down theirs. The Jews defended their fortress with valour, and during the siege various battles were fought, distinguished by traits of individual heroism. After a brave resistance of twenty-five days, the garrison surrendered at discretion. Seven hundred of them were dragged in chains to the public market-place of Medina, where a pit was dug to serve as a common grave, into which they were precipitated one after another, before the bloody hand of the executioner had time to extinguish the vital spark. This butchery of his helpless enemies the victor beheld with an inflexible eye, and makes the Koran applaud the Divine goodness in giving him the lands of the slaughtered idolaters as an inheritance.

The sixth year of the Hejira was distinguished by no other military events than a series of petty excursions, which added considerably to the wealth of the believers, and ended in the subjugation of several tribes of the Arabs. Zaid undertook an expedition into the territory of Midian, the same pastoral tract where Moses kept the flocks of Jethro. The adventurers were rewarded by a very considerable booty; besides a great number of women and children, whom they sold for slaves. Hitherto it had been customary in this inhuman traffic to dispose separately of mothers and children; but on this occasion, the cries and wailings of the female captives were so distressing, that the apostle produced a revelation prohibiting children of a tender age to be sold, except conjointly with the mother.

Successful in war, and enriched by conquest Mohammed saw himself not only at the head of a religious sect, but the sovereign of a petty kingdom. One acquisition yet remained, without which his authority, whether secular or sacred, could never be said to rest on a solid foundation. Mecca, the ancient and venerable sanctuary of Arabian worship, though he had granted protection to its commerce by an order to allow the caravans to pass unmolested, was still in the hands of his enemies, and a stranger to the true religion. To revisit the city and the temple from which he had been driven as a seditious outcast, and to which his followers still looked with a longing affection, was the cherished object of his fondest hopes: and the apostle imagined the time had arrived when he could gratify the devotion of his subjects; for such was the pretext under which he covered his design to surprise the capital of the Koreish. The Mussulmans were transported with delight; the holy banner was unfurled, and the most splendid preparations made for this famous expedition. Fourteen hundred of his bravest troops attended his march, ostensibly to protect him from insult or opposition. Seventy camels, adorned with garlands of flowers and leaves, the victims destined for sacrifice, advanced in front of the army.

This religious pageant did not, however, impose upon the inhabitants, who distrusted the intentions of an ambitious fanatic, veiled as they were under the humble garb of piety. Notwithstanding his declarations of peace, and of his extreme veneration for their temple, the Meccans informed him that if he entered their city it must be by force. "The Koreish," said Arwa, one of the deputies who had discoursed familiarly with the Prophet, "have put on their tiger-skins, and vowed resistance in the face of heaven." Mohammed sought a pretext for war and vengeance; and the conduct of his enemies in violating the law of nations, by seizing the person of Othman his ambassador, gave him the advantage of having the appearance of justice and right on his side. But neither party seemed disposed to appeal to the sword.

The Koreish foresaw the danger of wantonly provoking hostilities. They restored Othman to liberty, and sent a commissioned agent, Sohail, to propose conditions of peace. The preliminary words of the treaty were offensive to the idolaters; and in consenting to waive his usual title, Mohammed displayed weakness rather than policy. "In the name of God and his apostle," was the formula with which Ali commenced. Sohail remonstrated: "Had we acknowledged you to be the apostle of God," said he, "we had offered you no resistance" Ali was requested to efface the obnoxious words, and substitute, "the son of Abdallah." "No, by God!" cried this first of believers, "I shall never obliterate that glorious title. How should I be guilty of such profanation?" The Prophet, less scrupulous, with his own hand (and the assistance of a miracle) removed the objection, by writing his simple designation.

The conclusion of this truce filled the Mussulmans with shame and sorrow. They had left Medina on the faith of assured victory, and compassing the sacred enclosure of the temple; but, instead of triumph, their visit had ended in an ignominious peace. They could not conceal their chagrin, and for the first time the voice of their chief was disregarded. Part of the plain on which they were encamped lay within the sacred territory, and the apostle, willing to cover this disgrace with the solemnity of a pilgrimage, ordered his companions to slay their victims. The command was heard in mournful silence, and disobeyed. Three times the order was repeated, but the Moslems remained immoveable. The same presence of mind that had extricated him from other difficulties, did not fail him in this. Passing along the ranks without uttering a word, he seized the first camel, and with his own hand performed the rite of immolation. The force of example overcame their obstinacy. In an instant every victim was sacrificed, and every soldier occupied in the religious duties of shaving and purification, with a zeal and rivalry altogether marvellous. Their melancholy was entirely dissipated and harmony restored, by the descent of a new revelation, which assured them of speedy victory, though in the present enterprise they had anticipated the promised success.

The Jews were the doomed victims on which the fury of the rapacious believers was again to be let loose. Though weakened by exile and confiscation, several places of strength remained in their possession. With a body of 1400 infantry, and 200 horse, he directed his march to Khaibar, a fortress of prodigious strength, distant six days' journey to the north-east of Medina, and the capital of the Jewish Arabs. It was protected by eight castles, some of which were deemed impregnable. The besiegers opened their trenches; but all their assaults were vigorously repulsed. The gallant behaviour on both sides protracted the siege to a considerable length; and the Prophet was finally compelled to sound a retreat, in order to give his troops a few days relaxation.

This interval was signalized by some remarkable traits of individual courage. Abu Beker and Omar had successively mounted the breach, with a chosen detachment; but were forced to retire amid a shower of darts and arrows. The standard was committed to Ali, who fought with a valour more than human. In single combat he encountered Marhab, a gigantic Hebrew, governor of the castle, a man of prodigious strength and ferocity; and with a blow of his resistless sabre, called the piercer, he cleft him to the teeth, though his head was defended by a ponderous helmet, lined with a double turban. In the fray, the lance of his antagonist had struck his buckler to the ground; but the undaunted Mussulman supplied its place, by tearing from its hinges a gate of the fortress, which he wielded in his left hand during the whole assault, though the strength of eight men was found unable to move it from the spot where it lay. Such at least is the declaration of Abu Rafe, whose zeal for his master's glory, we cannot help thinking, has in this exploit rather overstepped even the modesty of romance. The fall of their champion dispirited the Jews; they fled in dismay to their castle, pursued by the victorious Moslems, who entered with the fugitives, and took possession of the fort.

This important conquest was followed by the surrender of the other castles; and lastly by the town of Khaibar itself, which being now destitute of its chief supports, was obliged at the end of ten days to capitulate. The conditions were humane, but mercenary; the inhabitants being permitted to cultivate their lands and vineyards as formerly, one-half of the future produce and of their present effects being awarded to the Mussulmans. This grant was coupled with a despotic restriction, that they held their possessions entirely at the will of the conqueror, who might expel them at his pleasure. Under these severe stipulations, the Jews continued to possess their castles and territories undisputed till the reign of Omar, who transplanted them to Syria; alleging the dying injunction of the apostle, that one religion only should be tolerated in his native land of Arabia.*

One event connected with this siege still remains to be noticed,—an attempt to poison the apostle while supping in the fortress with his chiefs on a

* According to Niebuhr, the Karaite Jews, in his time, were in possession of Khaibar, where they lived independent under their own sheiks.—*Descript. de l'Arabie*, p. 326. Burckhardt informs us that the Jewish colony once settled at Khaibar has wholly disappeared; nor are there any Jews in the northern part of the Arabian Desert. There are descendants of the Karaites still at Sanaa in Yemen. The Arabs of Khaibar are of a darker complexion than the neighbouring Bedouins.—*Travels in Arabia*, App. No. vi. p. 464

shoulder of roasted mutton. A single mouthful was sufficient to detect the fraud; but Bashar, one of his companions, having eaten heartily, was instantly seized with convulsions, and expired on the spot. The Arabs pretend that the mutton spoke and informed Mohammed of its being poisoned: this intelligence unfortunately came too late; for, notwithstanding the promptitude with which he rejected the deleterious morsel, he was persuaded till death that the malign influence had penetrated his system, and abridged his days. Revenge was the origin of this conspiracy; and the pretext, an experiment to try the reality of his apostleship. The perpetrator was Zainab, the sister of Marhab, who fell by the hand of Ali. "Had you been a true prophet," said the heroic Jewess, when asked the motives of her criminal intent, "the poison was harmless, as it must have been easily discovered; if not, it would have freed the world of a tyrant!"

Crowned with spoils and honours, the Prophet entered Medina in triumph. Confirmed in his kingly authority over his own subjects, he now assumed the insignia of royalty, and the prerogatives of an independent sovereign, in despatching agents and embassies to foreign courts to treat on matters of commerce, and especially to open their eyes to the precious benefits of becoming his disciples. Princes were not to be addressed in the ordinary style of epistolary correspondence; and Mohammed caused a silver seal to be made, on which was engraven, in three lines, MOHAMMED THE APOSTLE OF GOD. The vanity of his chiefs was flattered at the idea of becoming ambassadors; and seven of the neighbouring potentates, in the following order,—the King of Persia, the Emperor of the Greeks, the Governor of Egypt, the Nayash of Abyssinia, the King of Gassan, the Prince of Yemama, and the King of Bahrein, or Hira, were honoured with an apostolical invitation to embrace the faith of Islam. But the diplomatic missionaries were less successful with strangers than with the Arabs, who cherished a national reverence for the religion of Ishmael. The haughty Khoosroo tore the letter in pieces, because the name of "his slave," Mohammed, on the superscription, took precedence of his own. From the Emperor Heraclius, if we may believe the Mussulman writers, the sacred messenger had a more kindly reception; for they assure us, it was only the fear of losing his crown that prevented him from making an open profession of his belief in the Koran.*

With the functions of temporal sovereignty Mohammed conjoined that of chief priest or pontiff. During his life, he was himself the only minister and expounder of his religion. At first, such was the rude simplicity of the age, he used to preach in the mosque at Medina leaning upon a post, the trunk of a palm-tree driven into the ground. Accessions of power and magnificence required more appropriate accommodation; and at length he consented to have a stair or pulpit made, three steps in height, —the uppermost of which was occupied by himself; Abu Beker being seated on the second step; and Omar on the third, with his feet resting on the ground. Tradition asserts, that the first time the Prophet ascended the new rostrum, a dismal sound, like the lowing of a camel, issued from the deserted beam, expressive of grief and regret; and that the sympathizing apostle, caressing the disconsolate trunk in the most endearing language, restored it to good humour, and impressed it with a conviction of the propriety of their separation.

Nothing could exceed the respect and veneration in which Mohammed was held by his devoted fol-

* If we may credit Zonaras and other Greek writers, Mohammed had a personal interview with Heraclius, who was then at Emesa, on his return from his Persian expedition, and ceded to him a rich territory in Syria—*Memoires de l'Academ. des Inscript.* tome 32. *Gagnier*, lib v. chap. 4. *Abulfeda*, cap. 46.

lowers. His wishes were anticipated, his words and looks watched with the utmost attention. Every hair that dropped on the ground was gathered with superstitious care. His spittle was eagerly caught and preserved; and the water in which he had made his ablutions, as if it inherited a sacred virtue from his touch. The ceremonious expressions of allegiance, the formal servility of courts, are cold when compared with this fervour of a blind enthusiasm. "I have seen," said Arwa, the deputy of Mecca, who had contemplated the Moslem camp with leisurely astonishment, "the Khoosroos of Persia and the Cæsars of Rome in all their glory; but never did I behold a king among his subjects like Mohammed in the midst of his companions."

The eighth year of the Hejira was fortunate in the spontaneous conversion of three renowned proselytes, Othman, Khaled, and Amru, who most seasonably abandoned the sinking cause of idolatry. Othman was prefect or keeper of the temple, and the two others the future conquerors of Syria and Egypt. This same period was rendered memorable by the battle of Muta, the first in which the Moslems tried their swords against the disciplined valour of the Greeks. The apostle had sent an ambassador to the viceroy of Bosra, offering him as he had done to others the assurance of salvation on exchanging Christianity for the Koran; but the sacred messenger, while reposing in peace and security, was assassinated by the governor of Muta, a place opposite the town of Kerek, on the borders of Syria. This small spark kindled a vast conflagration, which overspread the East, and raged between the two nations for 800 years. An army of 3000 chosen troops, under the command of Zaid, was ordered to advance; and, on the spot where the guilty deed had been committed, to inflict on the perpetrator the chastisement of a just retribution. The burning sands of the desert were crossed by rapid and fatiguing marches; but the

Christians were not to be taken by surprise. A prodigious force of 100,000 men, composed of Greeks and auxiliary Arabs, was assembled. The Moslems hesitated whether to give battle or wait for reinforcements. "Friends," cried their chief, "let us cut our way to paradise through the ranks of the enemy. We have no alternative but martyrdom or victory!"

Fanaticism is blind to danger; and in spite of this vast disproportion of numbers, the believers flew to the attack, resolved to conquer or die. Seven years of triumph had inspired them with confidence in their good fortune, and a successful valour had made them invincible. The combat was long and bloody. Zaid fell among the foremost, covered with wounds and glory. The standard of Islam, as the Prophet had directed, was transferred to Jafar, brother of Ali, who more than sustained the fame of his country. When his right hand was struck off, he placed the sacred banner in his left; and when that too, was dismembered, he seized it between his bleeding stumps, and held it to his breast, though pierced with fifty wounds, till the blow of a Roman sabre cleft his head in twain. Abdallah, the third in command, met with the same fate; which so dispirited the hearts of the Mussulmans that they turned their backs for inglorious flight. At this crisis the intrepid Khaled raised the fallen standard, and succeeded in rallying the fugitives, who returned to the charge with redoubled fury. The centre of the enemy's line was broken, and thrown into complete disorder; the rout became general, and a terrible slaughter must have ensued had not darkness favoured their retreat. In a nocturnal council, the command was unanimously devolved on Khaled, who remained with his troops all night on the field of battle. No plunder had yet rewarded the bravery of the soldiers; but early next morning their new general, by a skilful manœuvre in deploying his ranks so as to magnify their numbers, contrived to spread a false apprehension among the vanquished Greeks, who fled with precipitation, leaving their camp, with the baggage and abundance of rich spoil, in the possession of the victors. This brilliant achievement added fresh laurels to the renown of Khaled. The Prophet bestowed on him the title of the Sword of God. To the science of an able captain he joined the most heroic personal courage; and during the battle, nine swords had broken in his hand.

The Greek writers speak less pompously of the battle of Muta than the romantic Arabs, and represent it as an action of no great importance to either side; but they seem to corroborate the leading fact that victory declared for the Moslems, who lost three emirs;* though their description of the imperial army, as "a body of troops hastily drawn together," would lead us to doubt of its numerical strength.

Two of the ten years' truce with the Meccans had scarcely elapsed when Mohammed accused them of a breach of their engagement. When a cause of quarrel is sought it is easy to find a pretext; but the truth is, that the condition of the contracting powers had somewhat changed. While his enemies were weakened by desertion and conquest, he had gained strength by the seduction or submission of various petty tribes. The opportunity was too favourable to be lost, as he had secretly determined to humble the pride of the idolaters and get possession of his native city. An army of 10,000 was ready to obey his will. Enthusiasm and revenge impelled their march; and, notwithstanding several attempts to give the enemy intimation, so completely were their

* Georgius Cedrenus (Historiar. Compend. p. 429), Theophanes (Chronograph. p. 278), mention the loss of three Arabian emirs, and the escape of Khaled, whom they call **Μαχαιραν του Θεου**

movements concealed, that the Meccans were only apprized of their arrival by the blaze of 10,000 fires within four parasangs of the city. Leaving Omar in charge of the camp to intercept all spies and communications, the Prophet advanced his army in four detachments to invest the town on every side. The left wing was conducted by Ali with the great standard of Islam at its head, and the right by Khaled; while he himself, mounted on his camel and clothed in a scarlet robe, took the rear, and expressly prohibited his generals from committing violence except in cases of defence.

The inhabitants were in a state of great consternation, but it was too late to offer effective resistance. Abu Sofian and two of their chiefs had fallen into the hands of Omar: to save their heads they swore fealty to Mohammed, not without suspicion of treachery.* The Moslem troops met with no opposition, and three divisions, without striking a blow, marched peaceably into the city just as the morning sun was appearing above the horizon. Khaled alone encountered a large body of the enemy, who disputed his passage through the plain with a shower of arrows; but, after a sharp conflict, they were speedily dispersed. Ignorantly or purposely disregarding the prohibition of the Prophet as to the effusion of blood, the furious chief pursued the fugitives into the heart of the town; massacring all that came in his way in the streets and public squares as far as the gate of the temple, where vast multitudes had taken shelter. Seventy are said to have fallen by his own hand; numbers sought the

* M. de Brequigny, in his excellent Dissertation (Mém. de l'Acad. des Inscriptions, tome xxxii.), conjectures that a secret treaty had been concocted between Mohammed and Abu Sofian at Medina, the preceding year. The marriage of the Prophet with his daughter Habiba, and the liberal share of spoil which he conferred on him and his two sons after the battle of Honain, certainly confirm the suspicion.

protection of their houses, while many fled to the hills to escape the carnage of the merciless barbarian.

The apostle himself, having caused the troops to desist from slaughter, entered not with military triumph, but in the humble guise of a pilgrim, with a black turban, and the *ihram* or sacred habit, repeating aloud the 48th chapter of the Koran. He rode a white camel, and was attended by a body-guard of his principal officers. Seven times he went in procession round the Kaaba, each time touching the black stone with the end of his cane in profound reverence; he then made his devotional inclinations, drank copiously of the Zemzem, and performed his sacred lustrations in a pail of that holy water; the rest of the Believers observing the same solemnities. The idols, the objects of his earliest and strongest indignation, were now within his reach, and everywhere presenting their hideous forms before his eyes. With his staff he struck a wooden pigeon to the ground, and broke it to pieces; and touching with the same implement all the images within the enclosure, he gave the signal for their demolition. "Curse your idolatries! what have our pious forefathers, mortals like ourselves, to do with your sorceries and your sacrilegious worship?" And instantly Abraham and Ishmael were dragged from their pedestals. Hobal, with his hoary beard and his divining arrows, was laid prostrate. Mounted on the shoulders of the Prophet, Ali pulled down the idol of the Khozaites from the top of the Kaaba. Saints and angels, male and female, with the whole fantastic group of heathen divinities, were ignominiously swept from the place in one common ruin, until the pride of ancient paganism was brought low, and the temple cleansed from the accumulated vanities of 2000 years. Over the wrecks of their shattered deities the victor harangued the trembling idolaters on the folly of their senseless adorations recommended them to believe in one God; and pointed to his victories as a triumphant proof of his claims to the prophetic character.

The laws of conquest gave him the right to make the citizens his slaves; but his anger was directed more against the idols than the inhabitants of the country; and instead of indulging in a cruel retaliation, the generous exile forgave the guilt and appeased the factions of Mecca by restoring its political rights and sacred privileges. Interest more than clemency might dictate this compromise, for the people merited their freedom by submitting to his authority and professing his religion; and when the chiefs of the Koreish were summoned to his presence, and humbly demanded his mercy,—"What treatment but chains and bondage can you expect from the man you have wronged, and whom God has made your master?"—"That of a brother and a kinsman," said the suppliants. "Go, then; you are safe,—you are free." The only exception to this general amnesty was the execution of four criminals, who had rendered themselves personally obnoxious to the conqueror.

The whole inhabitants, male and female, with the haughty Abu Sofian and the ferocious Henda at their head, took the oath of fidelity and allegiance; and thus within eight years after his banishment the orphan son of Abdallah was enthroned as prince and prophet of his native city. Mecca was henceforth declared an inviolable sanctuary, where it was unlawful to commit bloodshed or cut down a tree. Its temple, instead of the promiscuous homage that formerly disgraced it, was to be shut for ever except to the partisans of the Koran; and a perpetual law was enacted, that no unbeliever should dare, under pain of death, to set his foot within the *haram*, or holy territory.

Fifteen days were spent in regulating the affairs of the Meccan government, and planning military expeditions for the further destruction of idolatry, which had several strongholds in other parts of Arabia. The suburban divinities were hurled in contempt from the neighbouring hills. But, while crowds of terrified image-worshippers were making their submission, an obstinate remnant had resolved to maintain their ancient liberties against the arms and the eloquence of the Prophet. The tribes of Hawazan and Thakif, with their allies the revolted Saadites, had assembled a force of 4000 men under the command of Malec, for the protection of their gods; and having posted themselves at Honain, near Taïf, were determined to intercept the Moslem army on its return from Mecca. At the head of 12,000 brave troops Mohammed reached the valley towards evening, and found the enemy drawn up in order of battle, a narrow mountainous defile lying between them. Before sunrise he had made the requisite disposition for attack, displaying with secret pride the banners of Mecca and Medina; the latter of which was followed by 10,000, and the other by 2000 warriors.

The general of the enemy, profiting by his advantageous situation, had contrived to supply by stratagem what he wanted in strength. An unexpected attack threw the ranks of the believers into complete disorder. They fled with the utmost precipitation, and a moment of panic had well-nigh lost the fruits of fifty battles. The Prophet himself, with ten of his faithful companions, were all that kept the field. Stung with shame and disappointment he attempted to spur his mule into the thickest of the enemy in search of an honourable death, but his brave comrades interposed, three of whom fell and expired at his side. While the valley resounded with the bitter lamentations of their general, the flying battalions were persuaded to return, and the combat was renewed with fresh vigour. "The furnace is again rekindled!" observed the delighted apostle; and throwing a handful of dust to encourage his soldiers, he soon beheld the tide of victory change in his favour. The infuriated troops, with the imaginary aid of 16,000 angels, inflicted a merciless revenge on the authors of their disgrace.

Malec, with the wreck of his army, had retired to Taïf and shut himself up in that fortress, resolved to defend it to the last extremity. This was a place of too great importance to remain in the hands of an enemy; and from the vale of Honain, Mohammed marched without delay to put an end to the War of Idolatry, by effecting its reduction. Twenty days were wasted in useless operations, and with reluctance he saw himself compelled to raise the siege; but he retreated with an idle threat to return, exhorting his companions to trust in his never-failing source of angelic assistance. A voluntary submission, however, rendered a second attempt unnecessary. The terrified Thakifites expressed their readiness to profess Islam on condition that their goddess, Al Lattah, was preserved for three years, and themselves exempted from the obligation of prayer. But the Prophet was inexorable. Religion without prayer, he told them, was worthless; and as for their idolatry, he could not tolerate it for an hour.

The booty which this expedition placed at his disposal amounted to 6000 captives, 24,000 camels, 40,000 sheep, and 4,000 ounces of silver. A halt of thirteen days was employed in distributing the plunder, for the impatience of his soldiers could no longer be restrained. Instead of reproaching the Koreish for their disaffection at Honain, he endeavoured to secure their attachment and silence their calumnies by a superior measure of liberality. Abu Sofian alone was presented with 300 camels and twenty ounces of silver,—a suspicion, had there been no other, of his private instrumentality to the conquest of Mecca. The same munificence was extended to his two sons and all the other chiefs of distinction, and the wavering faith of the new proselytes was confirmed by the lucrative religion of the Koran. Various gratuities were dispensed to several strangers belonging to other Arab tribes; and to conciliate the affections of the avaricious troops who were deprived of their prisoners, Mohammed was content to resign his fifth of the plunder, and wished, for their sake, that the cattle he had to bestow were as numerous as the trees in the province of Tehama.* The submission of the Koreish, to whom all Arabia looked with veneration as the genuine descendants of Ishmael, was a signal to those tribes who still remained hostile that resistance was useless. Most of them offered their voluntary homage, and the more refractory were glad to preserve their lives and effects by yielding a reluctant allegiance.

The sceptre of Mohammed was triumphant, and all the petty chiefs from the shores of the Red Sea to the Persian Gulf, acknowledged his regal and priestly supremacy. His arms were now powerful enough to attempt foreign conquests, and cope with the strength of neighbouring empires. The wealth and fertility of Syria had attracted his cupidity; and, under pretence of anticipating the warlike preparations of Heraclius, he resolved to march without delay into that province. An army of 30,000 troops was assembled, and a holy war solemnly proclaimed against the Romans. The harassed Moslems entered on this expedition with reluctance. It was the season of harvest, and a time of scarcity, when their labour was imperiously demanded in collecting their vintage. But in vain did they beg a dispensation, and urge their different excuses,—

* A difficulty with regard to married captives had startled the conscientious Moslems; but the casuistic knot was solved by the Koran, which pronounced it lawful for believers to make concubines of the wives of *infidel* husbands.—*Gagnier*, lib. vi. chap. 5.

the want of money, horses, and provisions,—their ripe fruits, and the scorching heat of summer. "Hell is much hotter!" said the indignant apostle; and without concealing from them the fatigues and obstacles they must surmount, he proceeded with determined intrepidity in the execution of his plan.

Painful and weary was the march of the distressed army. Ten men rode by turns on the same camel; and though the historians of this campaign have invented a copious shower in reply to the Prophet's supplications, the exhausted soldiers were reduced to the necessity of drinking the water from the belly of that useful animal. After ten days' journey in a burning desert, the Believers reposed by the waters and palm-groves of Tabuc, a town lying midway between Medina and Damascus. While encamped here, Mohammed was informed that the Roman army had retired, which put a stop to the prosecution of the war. The terror of his approach is ascribed as the cause of their sudden retreat; but though the valiant Arab declared himself satisfied with their peaceful intentions, it is more than probable that in the languid and discontented state of his own troops he declined to hazard his fame and his fortunes against the martial array of the Emperor of the East. An interval of twenty days was employed in subduing or receiving the allegiance of the neighbouring chiefs. Most of them being Christians were allowed, on the terms of an annual tribute the security of their persons, the property of their goods, the freedom of their trade, and the toleration of their worship. Among the petty princes that offered their personal obeisance at the camp of Tabuc was John, the Christian governor of Ailah, who compounded for liberty of conscience and protection to himself and his subjects by the yearly payment of 3000 pieces of gold. As a mark of his favour, the Prophet complimented him with a rich cloak or mantle, which descended to the caliphs, and is said to have been kept with religious care on account of its great virtue in curing diseases, until it fell into the hands of the Turkish sultan, Morad Khan, who ordered the precious relic to be enclosed in a chest of gold.*

Peace and submission reigned in Arabia, which now presented the singular spectacle of unity in faith and government. The five kings in Yemen had confessed that their eyes were opened to the true light, and consented to hold their crowns under the jurisdiction of the Moslem vicegerents. Ali continued at the head of an army to preach Islam to these happy regions; the inhabitants contributing with their own hands to the demolition of their altars and their gods. The next two years of the Hejira were allotted to the final adjustment of certain religious matters, and the reception of deputies and orators who flocked from all quarters to the court of Medina. While his lieutenants were thus saluted with respect in every province between the Indian and Mediterranean Seas, the ambassadors who knelt before the throne of the Prophet "outnumbered the dates that fall from the palm-tree in its maturity." They were received with condescension and kindness; and the year of embassies, the ninth of the Hejira, proclaimed the extraordinary concourse

* The substance of this instrument, the *Diploma Securitatis Ailensibus*, is to be found in Abulfeda (Vita, cap. lvii. p 125); Gagnier (La Vie, lib. vi. chap. 11); and Savary (Abrégé de la Vie, p. 195). It was published in Arabic and Latin by Gabriel Sionita, at Paris, in 1630. Bayle (Dict. art. Mahom.), Renaudot (Hist. Patriarch. Alexand. p. 169), and Abulfarage (Asseman. Biblioth. Orient.), admit its authenticity. Grotius and Hottinger (Hist. Orient. p. 237) doubt it; and Mosheim gives in to the opinion that it was a forgery of the Syrian and Arabian monks, to mitigate the severities of the Saracens.—*Ecclesiast. Hist.* Cent. vii. part ii. chap. 5. To the possession of the mantle in question, the Turks used to attribute the success of their arms; and believed that they owed their prosperity and the cure of all their maladies to their drinking the water in which it had been dipped.—*Savary*, note, p. 197.

which the fate of Mecca had attracted to acknowledge the power or supplicate the protection of the conqueror.

Various arrangements were put in operation to consolidate the strength of the infant monarchy. Officers were appointed to collect and superintend the ecclesiastical revenues; and the opprobrious name of tribute was exchanged for that of alms or oblations for the service of religion. All former edicts or treaties implying liberty, or exemptions in favour of infidels, were revoked. Some rites in the great annual solemnity were altered; the ihram was declared the essential dress of all pilgrims in future; and the indecent custom, which had prevailed in the Times of Ignorance, of performing the seven circuits naked, was abolished. To give the sanction of his own example, and furnish a model to the faithful in all succeeding ages for the exact and acceptable discharge of this ceremony, Mohammed determined to make a valedictory or farewell pilgrimage. The piety of the Moslem world was kindled, and a flock of 114,000 obsequious devotees accompanied his journey. His camp included all his wives, who, riding on camels, were enclosed within pavilions of embroidered silk; and was followed by an immense number of victims for sacrifice, crowned with garlands of flowers. Every spot where he halted and said his prayers became consecrated; and the manner in which he executed the various rites, from the cutting of his hair and nails to the solemn act of throwing stones at the devil, is still religiously observed by the hajjis of the present day.

The general tranquillity was at this time interrupted, but not endangered, by the rash pretensions of one or two adventurers whom the success of Mohammed had tempted to become his rivals. Moseilama, prince of Yemama, had made his submission, and professed the Moslem religion among the host of ambassadors that flocked to Medina

Sovereign of a considerable province, and beloved by his subjects, he conceived the project of commencing apostle on his own account; and scarcely had he returned to his capital when he renounced Islam, and began to assume the ensigns and prerogatives of a Divine messenger. Affecting to be equal in honour and dignity with the Prophet, he wrote him an epistle, modestly proposing that the earth should be divided between Moseilama and Mohammed, the two "apostles of God." The latter was too well confirmed in his empire to need an associate, or tolerate this "piece of unparelleled impudence." "Let Moseilama, the liar, know," was the reply, "that the earth belongs to the Lord, who will give the victory to his true servant."

Until his sixty-third year, Mohammed had sustained, with unabated vigour, the temporal and spiritual fatigues of his mission. The infirmities of age had not impaired his constitution, though his health had suffered a gradual decline. His mortal disease was a fever, of which he was seized in the house of Zainab, one of his wives, while giving directions to Osama to lead an expedition into Palestine to avenge the death of Zaid, who had earned the crown of martyrdom at the battle of Muta. Finding his malady increase, he requested to be conveyed to the mansion of his favourite Ayesha, whose tenderness might sooth his last moments. To her he expressed his serious conviction that he owed the cause of his distemper to the poisoned mutton at Khaibar. For three days he suffered the tortures of an intense and insupportable heat, which deprived him at intervals of the use of reason. This paroxysm was succeeded by a more favourable crisis, and he recovered so far as to officiate at prayers in the mosque. His audience were edified by a penitential acknowledgment of his willingness to make restitution to such as he might have unconsciously wronged. "If there be any man whom I have unjustly scourged, I offer my back to the lash of retaliation. If I have aspersed his reputation, let him proclaim my faults. If I have taken his money, or despoiled him of his goods, I am ready to give the little I possess to compensate his loss. Let my accuser make his demand; it is not my disposition to resent the claims of justice."—"Yes," exclaimed a voice from the crowd, "you owe me three drachms of silver." Mohammed immediately discharged the debt, and thanked his creditor for accusing him in this world rather than at the day of judgment.

To his latest hour, and amid sorrow and suffering, he continued to act the character of the Prophet; evincing at the closing scene of mortality the same remarkable fortitude and presence of mind that he had displayed on the field of battle. In one instance only did the violence of disease betray his wandering faculties into a momentary illusion, when he called for pen and ink, that he might write a book for the better instruction of his followers, and to consummate the work of revelation. The proposal was startling, and met with opposition, as the Koran was deemed sufficient: the chamber of sickness was disturbed by an unseasonable dispute, until the dying Prophet was forced to reprimand the indecent vehemence of his disciples. Unwilling that his attendants should witness the recurrence of his infirmities, he ordered all persons to be excluded from his apartment; and the last three days of his existence were spent in the exclusive society of Ayesha.

Tradition, which disfigured his life with romance, has left us to contemplate the circumstance of his death through a cloud of superstitious incense. If we are to place the slightest credit on the evidence of his only companion, he received more incontestable proofs to establish the truth of his mission at its termination than in any former period. Gabriel made regular visits of condolence and inquiry after

Medina and Mosque of the Prophet.

his health. The angel of death was not permitted to separate his soul from his body till he had respectfully solicited permission to enter the chamber. The request was granted, and the last office performed with all the deference of a servant to the command of his master. When the moment of his departure approached, his head was reclined on the lap of Ayesha; he fainted in the agony of pain, but recovering his spirits, and raising his eyes with a steady look towards the roof of the apartment, he uttered with a faltering voice the following broken and scarcely articulate expressions:—"O God!——pardon me——have pity——Yes,——receive me——among my fellow-citizens on high!" and immediately expired on a carpet spread on the floor. The particular year of his death has been disputed: but the best authors fix it to the 12th of Rebiah I., in the eleventh year of the Hejira, corresponding to the 17th of June A. D. 632.

The melancholy intelligence spread rapidly through the city. The frantic populace would not be convinced that they had been deprived of their apostle. They rushed with credulous dismay to the house of mourning. Disregarding the evidence of his senses, Omar unsheathed his scimitar, threatening to strike off the heads of the infidels who should dare to assert that their master was no more. A scene of tumult and confusion ensued, which retarded the interment for some days. The ferment was at length appeased by Al Abbas and Abu Beker, who produced the testimony of reason, and a text of the Koran, that the Prophet had actually tasted of death.

New disputes arose as to the place of sepulture; Mecca, Medina, and Jerusalem being each suggested by the contending parties. A saying ascribed to Mohammed decided the point: "That a prophet ought to be buried on the spot where he dies." A grave was dug, and the coffin deposited in a vault paved with bricks, beneath the floor of the apartment where he had breathed his last; and to this day the innumerable pilgrims of Mecca often turn aside from the road, to bow in voluntary devotion before the simple tomb of the Prophet. The multifarious duties of the hajji who performs this expedition—his ejaculations on perceiving the distant trees and spires—his alms, perfumes, and prostrations—his prayers and postures within the sacred enclosure of the mosque where the sepulchre stands, are copiously described by the Mussulman doctors.

For the personal appearance and private life of the apostle, we must rely on the Arabian writers, who dwell with fond and proud satisfaction on the graces and intellectual gifts with which nature had endowed him. He was of a middle stature, of a clear fair skin, and ruddy complexion. His head and features, though large, were well proportioned; he had a prominent forehead, large dark-brown eyes, and aquiline nose, and a thick bushy beard. His mouth, though rather wide, was handsomely formed, and adorned with teeth white as pearls, the upper row not closely set, but in regular order; which appeared when he smiled, and gave an agreeable expression to his countenance. He had a quick ear, and a fine sonorous voice. His dark eyebrows approached each other without meeting. His hair fell partly in ringlets about his temples, and partly hung down between his shoulders; to prevent whiteness, the supposed effect of Satanic influence, he stained it, as the Arabs often do still, of a shining reddish colour. His frame was muscular and compact, robust rather than corpulent. When he walked he carried a staff, in imitation of the other prophets, and had a singular affectation of being thought to resemble Abraham. The assertion of the Greeks and Christians that he was subject to epilepsy must be ascribed to ignorance or malice.

The flattery or the superstitious veneration of his followers has created several attributes that are either fictitious or impossible; we shall therefore leave out of our portrait the sweetness and nutritive qualities of his spittle—the faculty of vision from behind—his miraculous exemption from vermin, which would neither touch his garments nor taste his blood—the odorous exhalations of his arm pits—and the delicious perfumes that exuded from his body like drops of liquid coral. We must also pass in silence his miraculous skill in the sciences; his arsenal and wardrobe, with his two black and white standards, called the Eagle and the Sun; his horses, mules, asses, camels, sheep, goats, and other bestial, of which many remarkable anecdotes are related in oriental authors.

The private and moral character of Mohammed was checkered by a strangely inconsistent mixture of virtue and vice, of dignity and condescension. Though vested with the power and ensigns of royalty, he despised its pomp, and was careless of its luxuries. The familiarity which gained the hearts of the nobles, and endeared him to his companions, was extended to the meanest of the people, whose wishes and complaints he always listened to with patience. He even entertained them occasionally at his table, or shared with them their homely meal while seated on benches around the mosque. When not occupied in matters of graver importance, he threw aside the forms and restraints of official etiquette, and condescended to partake in the amusements or jocular conversation of his friends. At the head of his army he could maintain the stateliness and cold taciturnity of Wellington or Bonaparte; with his soldiers he could relax without losing his authority, mixing in their pastimes and pleasantries with that freedom which reminds us of the sportive freaks of Cromwell, whose character for military genius, fanaticism, and hypocrisy he in many points resembled.

He courted no distinction beyond others in food

or dress. Dates and water, or a sparing allowance of barley-bread, the abstemious diet of his country were his usual fare. Milk and honey to him were luxuries: when he ate, he sat cross-legged on the ground; and when he travelled he divided his scanty morsel with the valet, who generally rode behind him on the same animal. On solemn occasions he feasted his companions with rustic and hospitable plenty; but months sometimes elapsed without the comforts of fire or cookery being seen on his hearth. To finish this portrait of his humility, we learn that he was in the habit of performing the most humble and menial offices of the family. The lord of Arabia disdained not to mend his own shoes and his coarse woollen garment; he milked the sheep, kindled his fire, swept the floor, and served his guests at his own table. His liberality in bestowing alms bordered on extravagance, and often left him without money or provisions for the maintenance of his household. The sincerity of his ghostly injunctions to charity and benevolence was attested at his death by the exhausted state of his coffers.

His attention to the cares of the toilette was extraordinary in a person so immersed in devotion and conquest. Whenever he went to the mosque, or on a military expedition, he carried with him a vessel of odoriferous ointment, antimony for eye-paint, a comb, and a mirror; and in default of the latter, he would adjust his headdress by the reflection of water. The two things on earth in which he most delighted were women and perfumes; the fervour of his piety, he affirmed, was increased by these sensual pleasures; and he took care that his religion should make ample provision for their enjoyment. Nothing scandalized his Christian adversaries more than the freedom with which he indulged his conjugal propensities. From the laws which he imposed on others he claimed special exemption for himself; and in the gratification of his carnal desires he scrupled not to subject his character as an apostle, and even the pretended counsels of Heaven, to the imputation of weakness and inconsistency. The most public and criminal excesses were legalized by Divine revelation. The sanctity of the temple, the distinction of fast-days and holy places, he might, in compliance with demands of nature, violate with impunity. The barrier of prohibited degrees which confined his followers was no limit to his passion, claiming, as they did, a peculiar and exclusive license.

His seraglio, instead of the legal number of four, contained fifteen or seventeen (others say twenty-six) wives; and, what is singular, all widows, excepting only the daughter of Abu Beker. Next to Kadijah, his beloved Ayesha engrossed the greatest share of his confidence and attention. She had the reputation of being the most accomplished lady of her time; and long after his death she was revered as the Mother of the Faithful. Her youth and beauty maintained the ascendant in the harem; but her behaviour, if not criminal, was at least indiscreet. In the nocturnal march against the Mostalekites, the loss of a pearl necklace obliged her to dismount; and the train proceeded, unconsciously leaving her behind. Saffwan, an officer of the rearguard, found her unveiled and overcome with sleep; and conducted her in the morning on his own camel to the camp. The temper of Mohammed was inclined to jealousy, and his enemies gladly seized the opportunity of wounding his domestic honour. But the bitter tears of Ayesha, and her protestations of innocence, softened his anger. From this perplexing dilemma he was relieved by a Divine revelation, which assured him of her inviolable fidelity. The accusers were chastised, by the same authority, with eighty stripes; and a law was published that no woman should be condemned as guilty, unless the ocular evidence was adduced of four male witnesses His marriage with Zainab, the wife of his slave and adopted son Zaid, was the dictate of an illicit passion, and a shameful breach of the matrimonial law of forbidden affinities.

Besides his wives he had several concubines, the most noted of whom was the Egyptian Mary. Her charms were irresistible ; and, notwithstanding the prohibitions of the Koran, the apostle was too deeply enamoured to exercise the virtue of abstinence. To avoid the scandal, he had recourse to secret intrigue; but Haphsa surprised him in her own chamber with his favourite captive. He swore he would for ever renounce the possession of his mistress, and she promised silence and forgiveness. Both parties forgot their engagements—the harem was in a flame of jealousy and revenge; but Gabriel again interposed with a sura to absolve him from his oath; and exhorted him to enjoy the bounty of an indulgent and merciful Providence, without listening to the clamours of his wives. To chastise their loquacious indiscretion, Haphsa suffered a temporary divorce, and the rest were condemned to the penance of a solitary month; during which time the Prophet met with no obstruction in fulfilling the commands of the angel. At the end of thirty days he summoned them to his presence, reproached them for their disobedience, threatened them with eternal separation, both in this world and the next, and hinted at the possibility of supplying their places by others more faithful and devoted. This threat was the more appalling, as no woman whom he had once espoused was permitted to cherish the hope of a second marriage. The main argument by which his apologists excuse his sensualities was the hope of multiplying his descendants. Yet all the inmates of his harem were childless; and not a son survived to support the decline of his life, or uphold, after his demise, the dignities of priest and king. Of his eight children by Kadijah, Fatima alone lived to enjoy his paternal tenderness. She married Ali in the first year of the Hejira, and became the mother of an illustrious progeny.

The literary attainments of the Prophet, like many other parts of his character, have been made the subject of controversy. Adopting the authority of the Koran, and the unanimous testimony of Arabian authors, most historians have espoused the conclusion that he was totally unacquainted with literature—ignorant even of the elements of reading and writing. Instead of being ashamed of this defect, his followers gloried in it, as an evident proof of his divine mission. The constant boast of the Arabs was, that the Koran, whose elegance and sublimity were universally acknowledged, could never have been produced without the aid of celestial instruction, by a man destitute of the very rudiments of education. To evade its force, the Christian writers alleged that he was assisted in the compilation of his imposture by various associates, and that the pretended revelations of Gabriel were really the composition of certain private secretaries. A swordsmith at Mecca—two nameless Christians, who possessed copies of the Scriptures—Sergius, the Nestorian monk—and Abdallah, or Salman, a Persian Jew, are all enumerated as accomplices in this impious fabrication. The conjecture of secondary aid is, indeed, rendered probable from his own words. "I know they will say," he remarks (chap. xvi.), "that a man hath taught him the Koran; but he whom they presume to have taught him is a Persian by nation, and speaketh the Persian language. But the Koran is in the Arabic tongue."

We are, however, by no means satisfied that Mohammed was actually the "illiterate barbarian" that history represents him; and, in spite of Abulfeda, Gagnier, Reland, Sale, and Gibbon, we have a suspicion that his ignorance was more assumed than real—one of those plausible disguises which he employed to throw a veil of mystery over his proceedings. This we may perhaps infer from his extensive commercial intercourse with the polished cities of Syria, and from the fact, that at the time of his birth the use of letters must have been well known at Mecca. The seven poems suspended in the Kaaba, the decree of the Koreish, and other documents intended for public perusal, necessarily presuppose that the arts of reading and writing were neither uncommon nor extraordinary attainments. That Abu Taleb, Abu Beker, Ali, and many others of the first Moslems, were familiar with letters, are facts that none have ever disputed. Hence the fair presumption is, that Mohammed was not altogether uninstructed in those accomplishments which were possessed by numbers of his fellow-citizens in the same rank and the same profession. It is difficult for hypocrisy to be consistent; and, notwithstanding all his care and circumspection, the mask sometimes dropped off. If we are to credit his biographers, he wrote letters to several foreign princes; in his treaty with the Meccans he erased his apostolic title, and with his own hand substituted his family name; and in his last illness he demanded materials to record his final instructions to his people. The Arabs, indeed, ascribe the latter to the effect of delirium or disease; and the other they explain by the intervention of a miracle. This, however, is but an unsatisfactory explanation; and the evidence of these incidental facts seems to attest that there were moments when his pretended incapacity was forgotten, and when he not only expressed a wish to exercise, but actually practised, that very art of which he and his historians maintained his total ignorance.

That, to a certain extent, Mohammed was a benefactor to his nation cannot be disputed. Gross and absurd as is the whole system of Islam, it possessed many principles in common with the true religion; and is, doubtless in every respect, far preferable to the degrading and monstrous idolatry which formed the ancient and prevailing creed of Arabia. It was a wise and humane jurisprudence that forbade the infant slave to be separated from the mother; that abolished the immolation of children to idols; and the barbarous system of burying females alive. But the language of commendation can extend little farther than to the repeal of obnoxious usages. To the praise of a great or enlightened statesman Mohammed has no claim. That he was superior to the age in which he lived is evident from the success of his imposture; but nothing, except the prejudices of habit or education, could persuade any rational being of his merits as a legislator, beyond that of imbodying his loose and obscure institutes in a written form.

Admitting, to their full extent, his mental and intellectual qualifications, his character as a conqueror was deeply stained with the vices of Asiatic despotism. To a candid reviewer of his actions, it may appear that Mohammed was severe from policy rather than cruel by nature; but this can be no apology or extenuation of his guilt.* For the neces-

* Voltaire, in his "Tragedy of Mahomet," the plot of which embraces the truce and capitulation of Mecca, makes the Prophet "imagine and perpetrate the most horrid crimes." This play, which La Harpe calls a *chef-d'œuvre* of the French theatre, has made its hero a monster of cruelty and injustice, with the view of vilifying religion under the name of fanaticism. But it is at variance with the facts of history, and betrays a gross ignorance of Arabian character and manners The poet himself confesses that he is unsupported by truth, and roundly alleges, "que celui qui fait la guerre à sa patrie au nom de Dieu, est capable de tout."—*Œuvres de Voltaire*, tome xv. p. 282. *La Harpe, Cours de Littérature*, tome viii. p. 377. Colonel Vans Kennedy has published an able criticism on this tragedy, exposing its palpable deviations from history as well as from the principles of the drama; but his zeal against the literary sins of the infidel Frenchman has led him to take a much more favourable view of the

sity which usurpation creates, the usurper must be held responsible. That the stern Prophet was not insensible to the tender feelings of humanity is attested by unquestioned evidence. His tears mingled with the general lamentation for the warriors who fell at Muta; and over the neck of the daughter of his friend Zaid he wept the loss of his most faithful companion;—his disciples expressing their astonishment that earthly sympathies should dwell in the bosom of a messenger from Heaven.

If his inordinate ambition had been content with that pre-eminence to which it might have aspired without a crime,—had he been satisfied with the grand national object of a moral and religious reformation,—and employed his transcendent and commanding genius in civilizing his barbarous countrymen, and reclaiming them from their senseless superstition, without the impious pretensions of a Divine revelation,—his vices and defects, palpable as they were, might have been overlooked or forgotten amid the splendour of his victories; and he might have earned a proud rank among the distinguished friends and benefactors of mankind. But to those who judge of individual worth apart from the pomp and glare of constant triumph,—who investigate coolly the causes of a nation's prosperity, the fame of the Arabian Prophet will not stand the test either of private excellence or of public usefulness. Rude and imperfect as were the ethics of those times, his moral character shrinks with guilty apprehension even from his own standard of virtue; and our admiration for his astonishing talents and success is quickly lost in abhorrence of the cruel and profane purposes to which they became subservient.

character and religion of Mohammed than is warranted by the transactions of his life, or the benefits he conferred on his country.—*Transact. of the Lit. Soc. of Bombay*, vol. iii.

CHAPTER VII

THE KORAN.

The Koran—Its reputed Origin—Held in great Veneration by the Moslems—Its literary Merits—European Translations, Du Ryer's, Maracci's, Sale's, Savary's—Sources whence its Doctrines were borrowed—Its leading Articles of Faith—Angels and Jin or Genii—Examination of the Dead by Munkir and Nakir—Intermediate State of the Soul—The Resurrection—Signs that precede it—Ceremonies of the Final Judgment—The Judicial Balance—The Bridge Al Sirat—Torments of the Wicked—Luxuries and Enjoyments of the Happy State—Women not excluded from the Mohammedan Paradise—Predestination—Prayer—The Mohammedan Sabbath—Ablutions—Circumcision—Alms—Fasting—Festivals—Prohibitions as to Food, Intoxicating Liquors, and Games of Chance—Civil and Criminal Code of the Moslems—Laws respecting Marriage—Theft—Courts and Officers of Justice—Traditions—Mohammedan Sects—The Sonnees and Sheahs—Their Hatred of each other.

The Koran, as is well known, imbodies the pretended revelations of the Arabian Prophet.* It was delivered by its author, and is still received by his followers, as containing every information in the shape of precept and instruction necessary for the guidance and spiritual welfare of mankind. According to them it had an origin far more sublime than that of human invention, its substance being uncreated and eternal, co-existent with the essence of the Deity, and inscribed from everlasting with a pen of light on the Preserved Table in the seventh heaven.

* Koran is derived from the verb Karaa, to read, and means the book to be read.—*Sale, Prelim. Dis.* sect. 3.

Each parcel, as revealed by Gabriel during a period of twenty-three years, was carefully treasured up in the memories of the faithful, or committed to writing by amanuenses, who for want of more dignified materials wrote them on palm-leaves, skins of animals, and shoulder-bones of mutton,—a device practised by the ancient Arabs, who preserved their poems and works of imagination on these rude tablets, tied together on a string.

The first transcript of this divine volume was thrown in promiscuous detachments into a chest intrusted to the charge of Haphsa, next to Ayesha the most favoured of the apostle's wives. Two years after his death the originals were collected and published by his friend and successor Abu Beker, who took this method of rescuing them from the peril of being lost or forgotten. The volume was afterward revised, or perhaps rewritten, by the Caliph Othman, in the thirteenth year of the Hejira. This prince had observed a great disagreement in the manuscripts already extant, those of Irak differing from the Syrian; both, however, were superseded by the new copies, which were distributed over the several provinces of the empire; the old being burnt and suppressed. This amended edition of Othman is that read by the Moslems of the present day.

Like the Jews, the Moslems hold their sacred book in the most extraordinary veneration, and attribute to it many cabalistic virtues. They will not suffer it to be read or touched by any of a different persuasion; and if found in their possession the crime might be capital. They peruse it with great respect, never holding it below their girdles, and always qualifying themselves by first performing their legal ablutions. They swear by it, consult it on all occasions of moment, carry it with them to battle, and inscribe verses or passages from it on their banners and their garments, as they formerly did o[n] their coins. Its principal sentences, written on the walls of their mosques, remind them of their social and solemn duties. They bestow upon it the exalted epithets of the True Book, the Word of God, the Director of Men and Demons, the Quintessence of all Sacred Compositions, and not only the greatest miracle, but the spiritual treasury of 60,000 miracles. They have been at pains to compute the number of verses, words, and letters it contains; and even the different times each particular letter occurs. Of the seven ancient copies, the first reckoned 6000 verses, the second and fifth 6214, the third 6219, the fourth 6236, the sixth 6226, and the seventh 6225; but they agree in the common amount of 77,639 words and 323,015 letters.

After the example of the Masoretic rabbis, the learned Moslems have introduced vowel-points to ascertain the true meaning and pronunciation; which, without this adventitious light, must often appear obscure.* The most ancient manuscripts now known are on parchment, in the Cufic character; the modern are in the Niskhi, on paper curiously prepared from silk, and polished to the highest degree of beauty. Exemplars are to be found in every public library in Europe; but, as the Christians are prohibited the use of the Koran, most of these have been taken in battle, and some of them belonging to princes and persons of distinction. Copies of peculiar elegance were found among the spoils of Tippoo Sultan. That most admired for the character of its

* Like the Hebrew and Greek, the antiquity of accents or vowel-points in the Arabic has been much disputed. Hottinger (Clavis Script. p. 403) and Adler (Museum Cufic. Borgianum, p. 34–37) contend that the language was never without them; though their shape and position have occasionally varied. Gregory Sharpe (Dissert. on the Origin. Power of Letters, p. 87) maintains that the vowel-points were not in use till several years after the time of Mohammed. So likewise think the Turks, who give Ali the honour of the invention.—*Mill's Hist. of Muham.* chap. v. p. 281, note.

writing and embellishments was formerly the property of Soliman the Great, and is preserved in the Museum Kircherianum at Rome.

Of the literary merits of the Koran the Arabs speak in terms of rapture. The most esteemed doctors of the mosque pronounced its style to be inimitable,—more miraculous than the act of raising the dead. Whatever may be its defects as a work of genius or merit, it is universally allowed to be written with great elegance and purity of language. Sometimes, in imitation of the prophetic and Scripture phraseology, it rises above the ordinary strain, and magnificently paints the Almighty seated on his throne of clouds and darkness, and dispensing laws to the universe. Though written in prose, it is measured into chapters and verses like the Songs of Moses or the Psalms of David. The sentences have the soft cadence of poesy, and generally conclude in a long-continued chime, which often interrupts the sense and creates unnecessary repetition. But to an Arab, whose ear is delighted with the music of sounds, and whose ignorance is incapable of comparing the productions of human genius, this metrical charm was its principal commendation; and was in fact so devoutly esteemed, that they adopted it in their most elaborate compositions.

All European translators have felt and acknowledged the difficulty of transfusing into their versions a lively image of those verbal and ideal charms peculiar to the original. The translation of Andrew du Ryer, a Frenchman, published for the first time at Paris in 1647, long maintained the highest credit but it is very dull, tame, and tiresome; and in his frigid prose we look in vain for the glowing and figurative expressions of the Eastern muse. Some years afterward (in 1698) appeared at Padua the Latin edition of Father Lewis Maracci, the confessor of Pope Innocent XI., and professor of Arabic in the College of Wisdom at Rome. It was the result of forty years' labour, and contains, in two folio volumes, a life of Mohammed, a refutation of his religion, the Arabic text of the Koran, with his own translation, and a vast collection of notes;—a work of such prodigious erudition as to merit a place in the same niche with the toilsome researches of the Benedictine monks. The zealous father, however, was more skilled in oriental than in Christian literature; more intent on exposing the frailties and blasphemies of his author, than in weighing his character or his religion in the balance of impartial criticism. His knowledge must obtain for him the respect of his readers, but his mode of reasoning will frequently excite their ridicule. He is one of that numerous class of writers to be found among the ponderous shelves of Continental divinity, who make no distinction between form and substance; and he pours as great a torrent of learning and argument on the trivial as on the important parts of the Mohammedan code.*

George Sale has maintained in England the hon-

* Parts of the Koran have been edited by Erpenius, Golius, Zehendorfius, Clenardus, Ravius, Pfeifferus, and Danzius. The first edition of the entire work in the Arabic was published at Venice in 1530, by Paganini of Brescia; but the pope was so alarmed that the book was immediately condemned to the flames; copies of it are therefore extremely rare. Peter, abbot of Chuni, in the fourteenth century, ordered a Latin translati to be prepared, which was published by Bibliander in 1550. A complete edition of the Arabic Koran was published by Hinckleman, at Hamburgh, in 1684. *Reineccii, Hist. Alcoran*, sect. 8, 10. *Peignot, Diction. des Livres Condamnés au Feu*, p. 227. *Mill Hist. of Muham.* p. 285. Purchas, in his Pilgrimes, and Heylin in his Cosmography, have given the chief heads of the Koran in English. The French of Du Ryer was translated by Alexander Ross, who thought it necessary to premonish the reader of his danger by "a needful caveat," of which the following is the exordium:—"Good reader, the great Arabian impostor now at last, after a thousand years, is, by the way of France, arrived in England; and his Alcoran, or gallemaufry of errors,—a brat as deformed as the parent, and as full of heresies as a scald head is full of scurf,—has learned to speak English."—*Ricaut's Hist. of Ottom. Empire*, vol. ii. See also *Retrospect. Review*, vol. iii.

ourable character of her Asiatic scholarship. His translation of the Koran is that with which we are most familiar, and has received the approbation of every master of the Arabic tongue. His Preliminary Dissertations are a valuable mine of Arabian history. Perhaps the only fault in his version is its being given in the form of solid and compact prose instead of being separated, as in the original, into verses. Maracci attended to this division, but he rendered it word for word; and has often disguised the thoughts and idioms of the flowery prophet in the unseemly garb of a barbarous latinity. Savary has also preserved this distinction; but he has done more,—he has infused into his French translation much of the spirit and beauty of the original. Among the numerous oriental versions, those into the Persic and Turkish are held in the highest estimation. Yet, with all these advantages, it will be difficult to impregnate the mind of a "European infidel" with any sentiment approaching that enthusiasm of respect and veneration in which the Koran is held by its own believers. He will peruse with impatience the endless repetition of pious declamation, the incoherent rhapsody of fable and precept, of promises, threats, and admonitions, which seldom excite any definite feeling or idea,—which sometimes, as Gibbon justly remarks, crawls in the dust, and is sometimes lost in the clouds.

Its materials are entirely borrowed from the Jewish and Christian Scriptures, from the legends of the Talmudists, the Apocryphal gospels then current in the East, the traditions and fables of the Arabian and Persian mythology,—all heaped together without any fixed principle or visible connexion. When describing the various attributes of the Divine Being, whether physical or moral, it conveys no clearer notions of the inscrutable essence than we before possessed. It only re-echoes the language.

and feebly imitates the expressions, of the inspired penmen. But the copy is far below the great original, both in the propriety of its images and the force of its descriptions. Its brightest passages are lost in the blaze of the purer light; and its loftiest strains must yield to the sublime simplicity of the Book of Job, composed, in a remote age, in the same country and in the same language. Even the enchanting fiction of the Mohammedan paradise was no original invention. The Jews had planted the mansions of the blessed in the seventh heaven, and furnished them with beautiful gardens. The idea of the celestial sphere is taken from the Almagestum of Ptolemy, whose writings were translated into the Arabic tongue, and have continued for seventeen centuries to be deemed the true astronomical system by the greatest part of the Asiatic world. The black-eyed houris were the creatures of the Magi. The streams, trees, nymphs, exquisite viands, and rich vestments, appear to be almost literally copied from the furniture of the Hindoo abode of happiness. The glorious but allegorical city of the Apocalypse, formed of gold and precious stones, with its twelve gates, its waters of life, and fruits of healing virtue, were grossly interpreted into sensual enjoyments.

With the theology of the Koran the reader is already acquainted. Its fundamental articles are comprised in the celebrated theorem, or confession of the Moslem faith,—LA ALLAH IL ALLAH; MOHAMMED RESOUL ALLAH—"There is but one God; and Mohammed is the apostle of God." This creed may be termed pure Deism; founded, as it is, on the unity and indivisibility of the Divine nature.

The Mohammedans divide their religion into two distinct parts,—faith and practice; each embracing a variety of subordinate particulars. The former or doctrinal department, besides a belief in God, in his Scriptures, in his prophets, and in his absolute decrees of predestination both of good and evil inculcates a similar persuasion as to the existence and purity of angels, the resurrection of the body, and a general judgment. These comprehend the six great points of faith; most of which have already been brought under the reader's notice. Angels are conceived to be spiritual beings created of fire, which neither eat, drink, nor propagate their species. It is heresy to deny their existence, or to assert any distinction of sex among them. Various forms, offices, and occupations are assigned to them; but four are more honoured, and oftener mentioned than the rest, as being higher in the favour and confidence of the Almighty: Gabriel, the minister of revelations; Michael, the friend and protector of the Jews; Asrael, the messenger of death; and Israfael, who will sound the last trumpet at the resurrection. They also believe that two guardian angels, who are changed daily, attend on every man to witness and record his deeds. The fall of Eblis or the devil, for refusing to pay homage to Adam at the Divine command, is a doctrine of the Koran; as is the belief of an intermediate order of creatures, the Jin or Genii, of a grosser fabric than angels, requiring nourishment, marrying, subject to death, and responsible, like other mortals, for their conduct.

The resurrection and final judgment of the human race have been adorned by the Mohammedans with many legends and fanciful embellishments. The day is to be preceded and ushered in with vast solemnity. Every corpse, when laid in the grave, is supposed to be catechized by two examiners, Munkir and Nakir, black and livid angels of a terrible aspect, who order the dead man to sit upright, and answer their interrogatories as to the soundness of his faith If his replies are satisfactory, the body is suffered to rest in peace, and refreshed by the air of paradise; if not, his torture commences. He is beaten on the temples with iron mallets, gnawed and stung till he receives his final doom, by ninety-nine dragons with seven heads each; which some interpreters allegorize into the acute anguish of a guilty conscience. There are sects who reject this notion of a sepulchral examination; but the more orthodox receive it as founded on the express authority of their Prophet.

The dissolution of the body by the stroke of death is palpable to the senses; and the existence of its immortal part is consonant to the wishes of nature and the speculations of the soundest philosophy. There is, however, beyond this separation, an interval of doubt and darkness, which neither the wisdom nor the restless curiosity of man has been able to penetrate.

The learned Moslems have exercised their ingenuity and their fancy in describing the various occupations and abodes of the soul in this intermediate condition, which they call Al Barzakh; but their opinions seem better adapted to convince the illiterate than to satisfy the doubts of the skeptic or the philosopher. The souls of the faithful they distinguish into three classes. Those of the prophets only have immediate admission into paradise. The martyrs undergo a sort of Pythagorean imprisonment in the gizzards of green birds, which feed on the fruits and drink of the waters of those delectable habitations. As to the disposal of the third class, there is not the same agreement. Some fix their abode near the sepulchres of their earthly companions; some with Adam in the lowest heavens; others enclose them in the trumpet of the archangel, or in the Zemzem well; while infidels are to be shut up in a certain pit in the province of Hadramaut; or hurled by the angels down to a dungeon in the seventh earth, under a green rock called the Devil's Jaw, there to suffer torment till soul and body are again united at the resurrection. Such are the idle and puerile discussions that have engaged the doctors of oriental theology; fables better suited for the amusement of children than for the academy or the mosque.

The precise time of the resurrection, Mohammed, wiser than many Christian fanatics, has not presumed to determine. It is a secret known to God only; and even Gabriel, when interrogated on that point, did not blush to confess his ignorance. But signs, both in heaven and earth, will darkly announce that awful catastrophe, when the material universe shall be destroyed, and the order of creation confounded in the primitive chaos. Eight lesser and seventeen greater phenomena will precede this general dissolution. The faith of men shall decay; the meanest shall be advanced to eminent dignities; and the maid-servant shall become the mother of her master. Among the greater signs are enumerated the sun's rising in the west; an eclipse of the moon, the appearance of a huge beast compounded of various species, more grotesque than the horned monster of the Apocalypse, which shall imprint her mark on the faces of all mankind, and demonstrate in the Arabic tongue the vanity of every religion except Islam; the coming of Antichrist, and the irruption of Gog and Magog, whose vast armies, in their career of slaughter and desolation, will drink the rivers dry, and whose bows, arrows, and quivers will serve the believers seven years to burn; the descent of Jesus on earth near the White Tower of Damascus, under whose reign the nations shall enjoy security and abundance. The relapse of the Arabs to their ancient idolatries,—the demolition of the Kaaba,—the speaking of beasts, birds, and inanimate things,—a smoke enveloping the whole earth, and a wind that shall sweep away the souls of all who have but one grain of faith in their hearts,—sum up the catalogue of these indications.

Still the hour of resurrection is left uncertain Three blasts of the trumpet (the Koran mentions only two, but the more orthodox sects have added another) will be the immediate signal, "when the whole earth shall be but a handful to the Almighty; and the heavens shall be rolled together in his right hand." The first trumpet, called the Blast of Consternation, will strike all creatures with terror,—shake the earth to its centre,—level the mountains,—darken the sun,—unsphere the stars,—and dry up the sea. The second, the Blast of Extermination, is the dread harbinger of death to all living beings,—a fate from which Azrael himself will not be exempted. After a pause of forty years will be sounded the Blast of Resurrection, when the dispersed particles of humanity, even to the very hairs, shall be re-assembled, and the souls imprisoned in the trumpet shall fly forth like bees to meet their respective bodies, filling the vast space between earth and heaven. This awful summons will recall to life every creature,—angels, genii, men, and animals; but the manner of their resurrection will be different. The destined partakers of eternal happiness will rise in honour and security,—those doomed to misery in disgrace, and under terrible apprehensions. The first-fruits of the grave will be Mohammed himself. His retinue will consist of three classes,—believers not distinguished for good works will march on first; those remarkable for piety will ride on white-winged camels, standing ready by their sepulchres, with saddles of gold or silver; the ungodly, timid and abashed, will creep grovelling with their faces on the ground; or, according to certain traditions, will change their shape into that of some brute typical of their vile propensities. While apes, swine, and intolerable stench designate respectively the sensualist, the miser, and the idolater; the unjust judges shall grope in blindness,—the false accusers gnaw their tongues in despair,—and the vainglorious be dressed in garments polluted with pitch.

The reunion of the soul and body will be followed by the final judgment of mankind; when each must give an account, and receive the reward of his actions. Though the procedure will he rapid and decisive,—not longer, as the Arabs express it, than the milking of a ewe,—a considerable pause will take place before its commencement,—a pause of anxious suspense both to the just and the unjust; the latter having their faces covered with blackness, and bathed in an agony of sweat, some to the ankles, some to the knees, and others to the mouth, in proportion to their several demerits; and this excessive distillation is the less surprising, considering the trampling and pressure of so vast a concourse, and the fiery beams of the sun, which will approach them within the length of a bodkin.

In his picture or copy of this sublime solemnity, Mohammed has too literally represented the forms, and even the slow and successive operations of an earthly tribunal. Each individual shall be minutely examined as to the circumstances of his life, and required to make public confession how he spent his time, or accumulated his wealth, or employed his talents and his learning. To these interrogatories he will be at liberty to offer the best defence in his power, by implicating others as the authors or partakers of his guilt. Even soul and body may dispute their respective share of criminality; and have the measure of their blame and punishment determined by the degree of evidence they can produce. The offending member may accuse the eye, and both plead in mitigation the carnal desire. The most exact measure of justice will be observed; and the good and evil deeds of mankind accurately weighed in a balance, real or allegorical, of so vast a capacity that its two scales,—one of which hangs over paradise, and the other over hell,—are large enough to contain both heaven and earth.* Into

* The believers in a literal balance allege that, as thoughts

these, thoughts, words, and actions shall be impartially cast, and according as the tremendous beam preponderates sentence will be awarded.

A singular mode of compensation will be allowed for the redress of injuries, but curiously illustrative of the Arabian doctrine of revenge. The aggressor must refund an equivalent of his own good deeds for the benefit of the person he has wronged,—the only means of reparation in his power. Should the balance still be in his favour, even to the weight of an ant, this remnant will secure his admission into paradise. But, on the contrary, should his stock of good works be exhausted, and any sufferers left who have not received satisfaction, his demerits will be burdened with an equal quantity of their sins, and the punishment due to them be visited on his guilty head. While the infidel part of mankind are condemned for their opinions, the actions alone of the Moslems will be examined; for their religious tenets, as the very name implies, are regarded as unexceptionably orthodox. The same rule of judgment will apply to genii and irrational animals; for both are held accountable. The weaker cattle shall take vengeance on the strong, and the unarmed on the horned, until the injured have entire satisfaction according to the strict law of retaliation; and when their wrongs are equitably adjusted, they shall be changed into dust,—the only exception to this doom of the brute creation being Borak, Ezra's ass, and

and actions have no specific gravity, the books in which these are written will be thrown into the scale.—*Sale, Prelim. Diss.* sect. iv. The idea of men's actions being recorded in a book is Scriptural. Matthew Paris, in the Vision of Thurcillus, has described the ceremony of weighing souls in the presence of the devil and the apostle Paul. Archbishop Turpin relates that, on balancing the merits of Charlemagne, the chance of salvation went against the emperor, until St. Jago threw into the scale the timber and stones of the churches which he had founded. This decided the matter; and the devil slunk off in rage and disappointment.—*Hist. do Imperad Carlos Mag.*

the dog of the Seven Sleepers, which, by special favour, will be permitted to rank among the true believers. The length of time consumed in the day of judgment, the Koran, in one place, makes 1000, and in another 50,000 years

Another perilous trial, which awaits every soul without distinction, is the passage of the famous bridge, Al Sirat (or the strait), which spans the dreadful abyss of hell, and is represented to be finer than a hair, and sharper than the edge of a sword This frightful path is beset with briers and thorns; but the good will find no impediment; they will cross with ease and safety,—Mohammed and Fatima leading the way; all the faithful being commanded to hold down their heads till she pass. To the wicked these obstacles will prove fatal; involved in darkness and dismay, they miss the narrow footing, and plunge into the fathomless gulf that yawns beneath them.

The regions of happiness and misery have been already partially described. It was the policy of Mohammed to terrify his followers by pictures of the most appalling torments; and regular degrees of suffering are ordained for every modification of guilt. The dark mansions of the Christians, Jews, Sabæans, Magians, and idolaters are sunk below each other with increasing horrors, in the order of their names; while the seventh or lowest hell is, with laudable justice, reserved for the faithless hypocrites and nominal professors of every religious system. This dismal receptacle, full of smoke and darkness, tradition asserts, will be dragged forward with roaring noise and fury by 70,000 halters, each hauled by 70,000 angels. The unhappy wretches will suffer from the extremes of heat and cold,—from the hissing of numerous reptiles, and the scourges of hideous demons, whose pastime is cruelty and pain. Despair will increase their misery; for the Koran has condemned them to these everlasting abodes without the smallest hope of deliverance. This eternity of damnation, however, is reserved for infidels alone, for the Prophet has judiciously promised that all his disciples, whatever be their sins, shall be ultimately saved by their own faith and his intercession. When the wicked Moslems drop from the narrow bridge, they fall only into the uppermost and mildest of the seven hells. The term of their expiation will vary from 900 to 7000 years; at the end of which, when "the crimes done in their days of nature are purged away," and their skin burnt black, they will be released; the infernal soot and filth being washed off in the river of life till their bodies become whiter than pearls.

With the structure of the Mohammedan paradise, and the peculiar nature of its felicities, the reader is somewhat acquainted. Before entering, the believers will be refreshed by drinking at the pond of their Prophet, which is supplied by two pipes from Al Kawther, one of the celestial rivers. Distinctions will be observed as to the time of admission; Mohammed, who will enter first, having declared that the poor (forming the majority of its inhabitants) will gain admittance 500 years before the rich. At the gate each person will be saluted by those beautiful youths appointed to serve and wait upon him, who will be the heralds of his arrival.

The measure of felicity will be proportioned to the deserts of the individual,—the most eminent degree being reserved for the prophets; the second for the doctors and teachers of the mosques; the next for the martyrs; and the last for the common herd of believers. The celestial joys of Mohammed were addressed chiefly to the indulgence of luxury and appetite. Rivers of water, trees of gold, tents of rubies and emeralds, beds of musk, garments of the richest brocades, crowns set with pearls of matchless lustre, silken carpets, couches and pillows of delicate embroidery, are among the rare treasures provided for the gratification of the externa. senses Other entertainments are on a scale of similar magnificence. Whatever is subject to waste requires sustenance; and the hungry saints will find abundant supply in a loaf large as the whole globe, in the flesh of oxen, and in the livers of fishes (delicacies among the Arabs), one lobe of which will suffice 70,000 men. While eating, each will be served in golden dishes to the amount of 300, and waited on by as many attendants. Wine, forbidden in this life, will be freely allowed in the next; and may be drunk to excess without palling on the taste, or incurring the risk of intoxication. The Tooba, or tree of happiness, so large that the fleetest horse could not gallop in a hundred years from one end of its shadow to the other,—bearing dates, grapes, and all manner of fruits, of surprising bigness and inconceivable relish, will extend its loaded boughs to the couch of every believer, bending spontaneously to his hand, and inviting him to pluck of its vintage. And should his capricious desires incline, its branches will yield the flesh of birds or animals, dressed according to his wishes; while from its expanding blossoms will burst vestures of green silk, and beasts to ride on ready saddled, and adorned with costly trappings. That every sense may have a congenial gratification, the ear will be ravished with the melodious songs of angels and houris, with the vocal harmony of the trees, and the Æolian chime of the bells that hang on their branches, moved by the soft winds of heaven. When to this train of gorgeous and sensual luxury are added the seventy-two damsels (the portion of the humblest of the faithful), whose charms shall eclipse all other glories, whose complexions are bright as rubies, and whose eyes, resembling "pearls hidden in their shells," shall never wander to any but their husbands, we may form a tolerable conception of those delights to which the voluptuous Mussulman looks forward as his chief felicity in another world.

The most exquisite and artificial pleasures of this life become insipid from long possession, or superfluous from the limited capacities of their mortal owner. Mohammed has made provision against both contingencies. At whatever period believers may die on earth, in heaven they shall never exceed the potent and animated age of thirty. A moment of happiness will be prolonged to 1000 years, and the enjoyment will be enhanced by an increase of abilities to the extent of a hundred fold. Amid the endless varieties of flavour and fragrance their appetites will never cloy; and they shall be exempted from those troublesome secretions which nature has made so indispensable to the health and comfort of the human animal. The eye of the enraptured possessor, so strong will be its vision, will wander over his gardens and groves, and descry the beauties of his wives and his wealth at the distance of 1000 years' journey. Should any of the faithful desire children (for without the wish the end would be unaccomplished), the space of one hour will suffice for the birth and growth of a young believer sixty cubits high,—the alleged stature of Adam, and the standard height of paradise. Or should his fancy turn to the rustic pleasures of agriculture, a moment of time will see his luxuriant crops spring up and come to maturity.

Women, so conspicuous in the Koran as incentives to religious zeal, have, by some, not only been excluded from heaven, but deprived of the attribute of immortality—a soul. This opprobrium must be ascribed to the misrepresentation of the Christians, or the ignorance of the Mohammedans, since its contradiction may be clearly inferred from their creed. Notwithstanding the prevalence of this vulgar error, the gates of paradise will be open to both sexes; but whether they shall inhabit the same or separate apartments is a point yet undecided. Mohammed had too much respect for the fair to teach such humiliating doctrine. His law rejected the negative precept of the gospel, of "neither marrying nor giving in marriage;" but he has prudently abstained from specifying the male companions of the female elect, whether they will be united to their earthly spouses, or have paramours of musk created for them; lest, as an ingenious historian has remarked, he should alarm the jealousies of their former husbands, or disturb their felicity by the suspicion of an everlasting union. Like men, their actions will be subjected to the same judicial balance, and rewarded or punished accordingly; though their felicity will not be so exquisite, as their deeds cannot have been equally important or meritorious. At the same time, the Arabian apostle has left it recorded, on the evidence of his own eyes, that the majority of infernal wretches are to consist of this frail and fascinating portion of humanity. Their degraded state in oriental society, as to their moral and intellectual character, may still be inferred from the legislative precept of the Koran, which estimates a woman as worth only the moiety of a man, and makes the fine of an offence against the former but one-half of that for an injury to the latter.

That the carnality of the future state, and an unlimited indulgence of the corporeal propensities, constitute a fundamental principle of the Mohammedan religion, numerous passages in the Koran place beyond all controversy. Some of the followers and modest apologists of that creed have revolted at so gross a doctrine, and adopted the convenient excuse of figures and allegories. But the sounder and more consistent party abjure the refined notions of types and parables; they adhere to the literal acceptation of the text, and would consider the highest metaphorical enjoyments a worthless substitute for the luxuries of the Tooba-tree, or the black eyes of the aromatic virgins.

Of predestination, as an article of Moslem belief, it is unnecessary to repeat what has already been stated. Its use, as a serviceable instrument in the hands of the warlike Prophet, and the effects it produced on his followers, was probably the only revelation that taught him the divine mystery. The apparent inconsistency between necessity and responsibility did not escape the penetration of his companions, who naturally reasoned, "Since God hath appointed our places, may we confide in this, and abandon our religious and moral duties?" But the son of Abdallah was not to be entrapped in this dilemma, and he replied, that good works were the spontaneous fruits of the happy, while bad were a characteristic test of the miserable. Over all the Mohammedan nations of the present day the tenet still reigns in its pristine force; and its effects are visible in that torpid inactivity of mind which supersedes the exercise of reason and industry, and considers every attempt to change the common order of things as a crime not far removed from rebellion against the established laws of the Deity.

The preceding sketch will suffice to give an idea of the first grand division of Islam, and the six cardinal points in its Confession of Faith. The second, or practical branch, comprehends four fundamental duties. 1. Prayer. 2. Alms. 3. Fastings. 4. The Pilgrimage to Mecca.—Prayer is a most important duty of the Mohammedans; it is declared to be the pillar of religion, and the key of paradise. The literal command of the Koran appears to enjoin only four times of daily prayer, called *namazi;* but a slight difference in the signification of the word has led the expounders of the sacred law to decide that five were meant: in the morning before sunrise,—directly after mid-day,—in the afternoon,—in the evening after sunset, while twilight remains,—and before the first watch, or midnight.* At these stated periods of devotion every true believer is summoned by the voice of the muezzins or public criers from the minarets,—*Allah akbar! Allah akbar!* &c. "God is great! God is great! there is but one God! Mohammed is his Prophet! Come to prayers! come to prayers!" and in the morning call are added the words, "Prayer is better than sleep! prayer is better than sleep!"—a sentiment not unworthy the consideration of those who profess a purer religion.

To indicate the direction of the Kaaba, towards which their faces must be turned, their mosques ave the *mehrab* or niche, pointing to the Holy City; and where these conveniences are impracticable, tables are furnished, calculated for finding the kebla. Their litany requires a certain number of ejaculations, which the more scrupulous count by a string of beads. Various ceremonies and attitudes are prescribed,—sitting, standing, kneeling, adoring with the face downwards, and seventeen *rikats*, or bowings of the body, two at morning prayers, three in the evening, and four at each of the other performances. In imitation of the old Jewish custom, or rather in consonance with the general feeling of Asiatic jealousy, the female sex are prohibited from joining in public prayers. Rejecting the days hallowed by the Jews and Christians, Mohammed consecrated Friday as the Sabbath of the Mussulmans.

* The words of the Koran are, "Until ye can distinguish a white thread from a black by the daybreak."—(Chap. ii.) The Jews determined the time of their morning lesson to be when they could discern the blue thread from the white in the fringes of their garments.—*Sale, Prelim. Diss.* sect. iv. p. 149. Mohammedanism is evidently not made for the Arctic regions. The Arabs divide their day into twenty-four hours, and reckon them from one setting sun to another. The moment when the sun disappears is called Mogreb; about two hours after is El Ascha; two hours later, El Marfa; midnight, Nus-el-Lejl; the dawn, El Fadjer; sunrise, El Subh; noon, El Duhr; three hours after noon, El Asr.—*Nieb. Descript. de l'Arabie.* The hours of prayer are, Mogreb, Nus-el-Lejl, El Fadjer, El Duhr, and El Asr

But its observance is by no means enforced with that strictness and decorum which distinguish the Jewish and Christian institutions. Except when engaged in the legal performance of their devotions, the faithful are allowed to resort to their pastimes or even their worldly employments. Originally, few mosques had regular preachers, the caliphs themselves discharging that function in a sort of harangue or exhortation to the people. The ancient Arabs appear to have had no other clergy than their chiefs or patriarchs. A priesthood or hierarchy of different ranks was afterward substituted over all the Moslem dominions; and perpetual revenues, from endowments or money gathered for pious uses, established for their maintenance.

Purification or cleansing of the body is an indispensable qualification, and pronounced by their apostle himself to be the key of prayer, without which it cannot be acceptable. The same ritual that prescribes their devotions enjoins on all Moslems a variety of preparatory legal washings. Certain cases require a total immersion of the body, called *ghasl*, which, on many occasions, is repeated three or four times a week. The *wodhu* is the ordinary ablution or washing of the face, hands, and feet, after a particular manner, and is necessary to the performance of every religious act. The number and repetition of these minute observances must to us appear idle and ridiculous; but the founder of Islam saw the utility of captivating the senses of a rude and ignorant multitude by a display of ceremonial rites. The precision with which they are described would often revolt the delicacy of European notions. Each ablution is regulated by the strictest order of method and precedency. Every stage of the process is accompanied with pious ejaculations:—the right hand has a prayer different from the left; the head, the neck, and ears, are washed each with its appropriate address to the Deity. The element may be rain river, sea, well, snow, or ice water; but it must be free of all impurities in taste, colour, or smell. When water is scarce or unattainable, there is permission to supply its place with sand, dust, gravel, ashes, &c This mode of performing the wodhu is called *Al taiamoum*. It has not, however, the merit of originality, for the Arabs and Jews made a similar use of earth, leaves, or buds of plants; and we read in a Greek author (Cedrenus) that the Christian rite of baptism was administered with sand to an expiring traveller in the wilderness of Africa.

Connected with this department of the Mohammedan liturgy is the ceremony of circumcision, both to males and females,—a rite in use among the idolatrous Arabs; as also in Egypt, Abyssinia, Ethiopia, and other countries to which the laws of the Jews did not extend. No certain age is prescribed, nor is the practice absolutely necessary: but the seventh year is esteemed the most convenient; and it is a reproach to all good Mussulmans to want this national symbol of their faith. It is always in the house of their relations that this religious ceremony is performed; the operators are the public barbers, and the patients are subjected to a confinement of fifteen days. The imam of the mosque assists in his priestly capacity; and, in families of distinction, the occasion is celebrated with alms, sacrifices, feasting, and other sumptuous entertainments. It is not to be confounded with the rite of baptism, which usually takes place on the day of the birth; the imam simply pronounces the name, and whispers into the child's ear an exhortation to be faithful to his creed, and attentive to prayers.

Charity is imposed as a religious duty on every Mussulman. The Koran frequently and strongly recommends it; and, in addition to what is exacted by law, every believer must make donations to the indigent, otherwise he is not considered to have performed the duty of almsgiving in all its extent

Alms are of two sorts, legal and voluntary. The former, by some called *zakat*, is of indispensable obligation, and defined by the legislator with minute precision, both as to kind and quantity. Originally, the zakat amounted to two and a half per cent. on the principal of the estate; and was commanded to be paid in cattle, including camels, cows, and sheep, money, corn, fruits, and even merchandise. Strictly to accomplish the law, every Mussulman was enjoined to bestow a tenth of his revenue; and if his conscience should accuse him of fraud or extortion, the tithe, by way of confiscatory atonement, was enlarged to a fifth. Upon the wide diffusion of Islam, the zakat was found to be not only difficult to collect, but unequal and invidious. The learned doctors split into a multiplicity of opinions about the proportion and mode in which it should be levied on property of various kinds. It has in consequence been generally abandoned, or restricted to goods imported by way of trade. The voluntary alms is left to the conscience of the giver: but the obligation of charity is so vehemently recommended, both by precept and tradition, that few orthodox believers evade the sacred duty.

The creed of Mohammed is hostile to the ascetic virtues. The voluntary penance of monks and hermits was odious to a prophet who censured in his companions a rash vow of abstaining from flesh, women, and sleep. The legitimate purpose of fasting, it is obvious, is the prevention of offences, not their punishment; but instead of making it a frequent or arbitrary observance, he has entirely frustrated its beneficial effects, by restricting it to a particular season of each year. Although voluntary penance is recommended, the month of Ramadan is specially set apart for religious abstinence. For thirty days, between the first appearance of the two new moons, the various members of the body must be kept under rigorous prohibition. To taste food or drink, to smell perfumes, or swallow spittle, to vomit, bathe, or even breathe the air too freely from daybreak till sunset, would render this sacred ordinance null and void. But from evening till daybreak, the faithful are allowed to refresh nature, though the more scrupulous renew their fast at midnight. When the Ramadan falls in summer, this self-denial is extremely rigorous and mortifying; for the patient martyr must wait the close of a tedious and sultry day, without assuaging his thirst with a drop of water, or tasting a particle of nourishment that can recruit his strength or gratify his senses. During this consecrated season other duties acquire an additional merit; charity becomes doubly virtuous, and the retaliation of injuries is forbidden. But, like other external ceremonies, this law is accounted a dead letter unless the performance is accompanied with a suitable disposition of heart and spirit. The only amends for these statutory mortifications, are the two *bairams* or principal annual festivals. The former (*Id al Fetr*, or feast of breaking the fast) begins on the first of the month immediately succeeding Ramadan, and is kept from three to five or six days. The other (*Id al Korban*, or feast of the sacrifice) commences on the 10th of Dulhajja, during the time of the pilgrimage.

The nature of oriental climates has rendered particular kinds of food detrimental to health, and led to a division of animals into clean and unclean. The filthiness of the hog, and its tendency to engender cutaneous diseases, have caused it to be proscribed in most warm countries. The interdict laid down in the Koran, in which the Mussulman doctors comprehend beasts and birds of prey, does not extend so far as the Mosaic catalogue. All amphibious animals are unclean; so are the ass and mule: but lawyers differ about the horse. Camels are lawful, hares neuter; but it is a mistake of ignorant writers to accuse the Arabs of feeding on dogs and wolves. The inhibition to eat blood was levelled against a common practice in the time of paganism, of drawing it from a live camel, which they cooked by boiling in a bag or gut. The animals proper for the diet of the faithful must be killed agreeably to a prescribed form; if slain accidentally, or in hunting, they may be eaten; but the most orthodox fashion is by cutting the throat.

Reasons both philosophical and medicinal have been urged for abolishing the drinking of wine. The first injunction of the Koran was simply against excess in the use of strong liquors, as incentives to quarrelling and bloodshed. This not proving sufficient, the special restriction was converted into a positive and general law, by which inebriating fluids were altogether proscribed. A precept so clear would hardly seem capable of admitting a latitude of interpretation, yet a diversity of sentiment exists as to its exact import; some contending that the statute allows the moderate use of wine, while the more conscientious hold it absolutely unlawful not only to taste that liquor, but to make it, or traffic in it, or even to maintain themselves by the money arising from its sale. The libertine and the hypocrite find means to evade the statute, and indulge freely, notwithstanding the threatened pains of hell.

The Moslem lawyers, with the casuistry of their profession, have so refined the simple injunctions of the Koran, that their essence is nearly lost. The crime of tippling may be practised with impunity to any extent short of outrageous disorder. If the smell of wine be not on the breath of the accused, or his intoxication self-apparent, evidence to the fact is of no avail; except in cases where the flavour may be presumed to evaporate, from the distance the offender has to travel to the residence of the magistrate. Even if the odour remain, or if he should vomit wine, witnesses must have seen him drink the forbidden juice; for, as the muftis and moollahs ingeniously argue, he may only have sat among wine-drinkers, or wine may have been administered to him by force or fraud. When the crime is fully proved, eighty stripes is the punishment of a free man; but a slave is liable only to forty, on the principle that, as bondage deprives him of half the blessings of life, he should suffer but half its punishments; all offences being supposed to increase in magnitude in proportion to the rights and enjoyments of the guilty. The inhibition against intoxicating liquors has been extended by the more orthodox to coffee, opium, tobacco, hashish, and benj, or the leaves of hemp in pills or conserve; but at present the whole of these articles are not merely tolerated, but used without any religious scruple whatever.

The moral argument against intemperance in drinking applied with equal force to the prohibition of gaming. Dice, cards, tables, sortilege, all Moslem commentators agree to be expressly prohibited. An artful and plausible distinction saved chess, the favourite pastime of the East, from this sweeping ordinance. That its success depends less on chance than skill and management has satisfied most of the Moslem nations of its lawfulness; who allow it under condition that it be not made a speculation for money, or a hinderance in the regular performance of their devotions. The fulminations of the Prophet are interpreted to have been directed chiefly against the carved pieces of ivory or wood which the idolatrous Arabs used in playing, being rude figures of men, elephants, horses, and dromedaries; and consequently condemned in the same text with image-worship. The pure orthodox sects substitute plain pieces of wood and ivory; but the Persians and Indians are less scrupulous about the sin of using carved images, or betting money.

The civil and criminal laws of the Mohammedans are based on the Koran, and extended into a sort of digest in various collections of supplementary traditions. To enter into the several decisions and interpretations of the more learned civilians might be curious, but would engross too large a space for our purpose. The restraints on polygamy, and the punishment of conjugal infidelity, have already been incidentally mentioned. The sole privilege which the laws of the Prophet give to a wife, but deny to a concubine, is that of dowry. In point of reputation their characters are without distinction; and the children of both are held legally in the same consideration. The power of dissolving the nuptial bond by divorce is granted to both sexes, but with different privileges. By the Mohammedan law, as well as the Mosaic, reasons which to us appear trivial justify the husband in severing the tenderest of all human connexions. Though a written divorce with the Moslems is in general use, the verbal declaration of the husband is sufficient; but this nominal facility of repudiation is powerfully checked by subsequent circumstances. In securing to the widow and the orphan that part of the property which is settled on her at marriage, and which, in the times of paganism, was often unjustly taken from them, the author of their system has vindicated the right of the female sex, and entitled himself, in this instance, to the praise of a wise and humane legislator. A woman's dower is entirely at her own disposal, and totally free from the control of her husband or his creditors; and so great is her independence in this respect, that mothers frequently assign the reversion to their sons, who have compelled payment of it from their fathers,—a privilege in singular opposition to the general condition of females in Eastern countries.

The fourth chapter of the Koran has detailed at length the laws relating to infancy, succession, and dowry,—matters of high importance in the estimation of the Moslems; and, in the fifth, the power of testamentary disposition is acknowledged, and several directions given for making this will in a solemn and authentic manner. The Mohammedan punishment of theft is unreasonable, and breathes the spirit of a barbarous age. "If a man or woman steal," says the Koran, "cut off their hands, in retribution for what they have committed."* According to the practice of the best Moslem courts, amputation is not to be inflicted unless the value of the stolen articles amounts to five dinars, or forty shillings,† and unless they be found in custody. In all cases where the penalty attaches, the right hand is to be struck off for the first offence, and the left hand for the second. Incarceration is the punishment for reiterated offences. The whole of a band of robbers are answerable for the acts of any one of their number; and if murder is also committed by them on the highway, they are put to death, even if the blood-avenger should forgive them; the sentence may be crucifixion or amputation, or both, at the discretion of the judge.

With regard to injuries or crimes of an inferior nature, where no particular punishment is awarded by the Koran, and which are not expiable by fine or compensation, the Mohammedans, as the Jews did in similar cases, have recourse to the bastinado, the most common chastisement used in the East at the present day; and which is performed by beating the offender on the soles of his feet. As they are fond of a celestial origin for their civil as well as sacred institutions, the cudgel, the instrument generally

* In cases of theft, which are punished by amputation, slaves and freemen are on an equal footing; for the Moslem doctors have gravely decided that it is impossible to halve amputation.—*Mill's Hist.* p. 325. D'Ohsson (Tab. de l'Emp. Ott.), Reland (De Relig. Mahom.), Bobovius (De Litur. Turc.), will explain the laws and liturgy of the Moslems.

† Five dinars may be reckoned equal to about 2*l.* 6*s.* 3*d.*, at the present value of money.

employed to carry the judge's sentence into execution, and which has been found of such admirable efficacy in keeping the naughty in good order, they pretend to have descended from heaven.

As their religious and juridical code is the same, the clergy are expounders of the law. Three general classes of judicial officers are recognised,—muftis, cadis, and mushtahids; but their powers vary in different nations. In India the cadi is the supreme civil judge. In Turkey he is the ordinary judge. The mufti is the nominal chief magistrate; but he has no tribunal, and never decides causes except those of great moment. The cadi is the officer who gives the law operation and effect; and in all questions of importance he is assisted by several moollahs, or learned men. The Koran, or its most esteemed commentators, regulate his decisions; and in novel cases he exercises his own judgment. The mosque is the place where he must sit for the execution of his office; or he may use his own house, so that there be free access to the people.

Besides the Koran, various traditions of the actions and sayings of the Arabian Prophet have been preserved; and these constitute the second authority of Mussulman law. The Koran was suited only for a rude people; and when the power of the caliphs increased, it became impossible to govern their numerous subjects by the comparatively few rules and maxims which it contained, most of which were local, and quite inapplicable to many of the nations who had submitted to the Saracen arms. This deficiency admitted only of one remedy,—that of rendering the original law more copious by authentic supplements. As the founder of their economy was believed never to have spoken but by inspiration, an account of these traditionary sayings was carefully noted down from his wives and companions. This immense collection was called *sonna*, a word equivalent to *custom* or *institute*, and somewhat resembles the Jewish Mishna.*

Of these collections six are held in peculiar esteem. One was made by Abu Horaira, a constant attendant on Mohammed, who nicknamed him the *Father of a Cat*,—an animal of which he was particularly fond, and carried always about with him. That formed by Abu Abdallah of Bokhara is peculiarly famous. Two hundred years after the Prophet's death he selected 7275 genuine traditions from 100,000 of a doubtful, and 200,000 more of a spurious character. This collection was compiled at Mecca and is adopted by the Sonnees.

Neither the simplicity of his creed, nor the terror of the sword, nor the exhortations of the pulpit could establish that unanimity of sentiment which Mohammed professed so ardently to desire. The spirit of division, which appeared among his followers even before his death, broke out with greater violence under his successors. The Mussulman sects have been far more numerous and violent than those of Christianity; and the history of the mosque presents as melancholy a view of the weakness of the human heart, and the pride of the human intellect, as is afforded by the annals of the church. Volumes might be filled with their names and their tenets. The same knotty points in scholastic theology that puzzled the divines of Christendom,—the

* The only complete work in the English language expressly on these traditionary laws, is a translation of the Mischat ul Masabih, or "Niche for holding the Lamp." The English version is by Captain Matthews of the Royal Artillery, and was published at Calcutta, in 2 vols. 4to, in 1809. The Mischat was properly a commentary on the Masabih ul Sunnat, or "Lamp of Religious Observances," collected by the Imam Hussein of Bagdad, who died A. H. 516. Another juridical commentary is the Hadaya, a work of very high authority in all Moslem countries where the Sonnee faith prevails. In 1791, Colonel Charles Hamilton published an English edition, in 4 vols. 4to. The first volume of Colonel Baillie's Digest of Mohammedan Law was published at Calcutta in 1805.

essence and attributes of God,—the justice of predestination, and its compatibility with freedom of will,—the province of reason in matters of faith,—and a number of casuistical questions on the moral beauty or turpitude of actions,—have been the theme of bitter and implacable controversy among the doctors of Islam.

The two leading sects are denominated the Sonnees and the Sheahs, or Shiites. The difference between them was originally more political than religious. The former appropriate to themselves the name of orthodox; they are traditionists, or believers in the Sonna; and, consequently, acknowledge the authority of the first caliphs, from whom most of these traditions were derived. Distracted with controversy, they at length reposed on the faith of four eminent theologians,—Hanifa, Malec, Shafei, and Hanbal,—who were not only reputed holy and learned divines, but masters in jurisprudence. These interpreters were in some points not altogether unanimous; but they were deemed radically sound, and have given their names to four sects, honoured as the pillars of the Sonnee faith. After their death each had a separate oratory in the Temple of Mecca, and they have been canonized as the four imams, or high-priests, of the established orthodox religion.*

The Sheahs, or sectaries, differed mainly from the preceding in asserting the divine and indefeasible right of Ali to succeed the Prophet. The same spiritual and secular dignity which they conceive should have descended immediately to the father, ought to have been transmitted to his lineal posterity; consequently, they consider, not only the three first caliphs, but all their successors, who took the title of Lords of the Faithful, as usurpers. This be-

* Hanifa was poisoned at Bagdad, A. H. 150. Malec died at Medina, A. H. 178–9. Shafei was a native of Syria, and ended his days in Egypt, A. H. 204. Hanbal died at Bagdad, A. H 241.—*D'Ohsson. Tab. Gen. Introd. Sale, Prelim. Diss.* sect. viii.

lief is hostile to the whole fabric of the Sonnee tradition, which rests on the authority of these three caliphs; though they admit the Sonna where its sources have not been contaminated. The Persians were the first nation who proclaimed themselves of this sect, about the commencement of the Suffavean dynasty (A. D. 1499); and for more than three centuries their creed has been the prevailing faith of that country. The authority of Hanifa maintains he ascendant in Turkey, Tartary, and Hindostan; Malec is chiefly recognised in Barbary and the southern parts of Africa; Shafei has followers both in Persia and Arabia, and possesses a limited influence over the seacoast of the Indian peninsula and the eastern islands.

There are numerous other heretical sects among the Mohammedans, who disagree even on fundamental points of faith. The spirit of hostility between most of these, especially the Sonnees and Sheahs, is rancorous and irreconcilable. Names which are never mentioned but with blessings by the one are hourly cursed by the other. No wars, as has been justly remarked, that ever desolated the Christian world, have caused half the bloodshed and misery, or been so deeply stamped with the character of implacable animosity, as have arisen from the political and religious controversies of the Mohammedan sectaries.

Having thus laid before the reader a sketch of the life and religion of the Arabian Prophet, we shall now advert to the warlike achievements of his followers, who constantly appealed to their victories as an express testimony of Heaven to the truth of their creed; and who, under the terrible name of Saracens,* extended their dominion over more kingdoms and countries in eighty years than the Romans had done in 800.

* The derivation of the name *Saracen* has puzzled etymologists. Some have suggested Sarah, the wife of Abraham; but they forget that the Ishmaelites were descended from Hagar

CHAPTER VIII.

CONQUESTS OF THE SARACENS.

Disputes in choosing a Successor to Mohammed—Abu Beker elected Caliph—Ali refuses Submission—Turbulent State of the Empire—Invasion of Syria—Success of the Saracens—Capture of Bosra—Siege of Damascus—Battle of Aiznadin—Surrender of Damascus—Death of Abu Beker—Accession of Omar—Pursuit and Plunder of the Damascene Exiles—Action at the Fair of Abyla—Battle of Yermouk—Siege and Capitulation of Jerusalem—Journey of the Caliph to that Capital—Surrender of Aleppo—The Castle besieged and taken by Stratagem—Reduction of Antioch—Flight of Heraclius—Subjugation of Syria and Palestine—Disgrace and Death of Khaled—Invasion of Persia—Battle of Cadesia—Occupation of Madayn—Immense Booty—Battle of Nahavund—Defeat and Death of Yezdijird—Final Conquest of Persia.

It was a political error in Mohammed, and one that proved fatal to the unity and stability of his empire, that he neglected to name his immediate successor, or lay down regulations for filling the vacant

Saraka, a city of the Nabathæans, *Sahara*, a desert, and an Arabic word signifying a thief, have all been adopted as the true etymon of the name.—*Stephan. de Urbibus. Hotting. Hist. Orient.* lib. i. cap. 1. *Bocharti Opera*, vol. i. col. 213. *Asseman, Biblioth. Orient.* tome iv. p. 567. *Abulfed. Geog. a Gagnier*, p. 63. The appellation has no allusion to any particular city, or any trait of national character. It comes from the Arabic word *Sharak*, and means an *Eastern People;* which the Saracens were in reference to the Romans.—*Pococke, Specim.* p. 33–35. Quid enim sonat Saracenus quam *Sharkion* et in plurali *Sharkiin, i. e.* Orientes incolas.—*Noble's Arab. Vocab.* p. 105. It was not till after the Roman conquests in Palestine that the name was known in Europe, when it superseded that of Ishmaelites and Nabathæans. It was obscurely applied by Pliny and Ptolemy to certain tribes, and used in a larger sense by Ammianus and Procopius. But it was not adopted as a national designation by the Christians until the year 715, in the reign of the Caliph Walid.—*Marigny, Hist. des Arab.* tome ii. p. 393.

caliphate. This is the more surprising, as he left no male posterity of his own, and must have foreseen the dismal consequences of an interregnum, or a disputed succession. His demise was the signal for immediate contest between the two grand parties of his followers. The same day that laid him in the grave saw them assembled to deliberate on the choice of a new sovereign. The Refugees insisted on their prior claim, as being the fellow-citizens, the kinsmen, and first proselytes of their apostle. The Ansars pleaded their meritorious services in offering an asylum to the fugitives of Mecca and their persecuted master. The Koreish were still jealous of the pre-eminence of the line of Hashem. The hereditary title of Ali was opposed by Ayesha, and offensive to the aristocratic spirit of the other chiefs, who were secretly anxious to keep the sceptre within their reach by a free and frequent election. Omar and Abu Beker were both proposed, but they mutually declined to take precedence of each other.

Separation appeared inevitable, swords were drawn, and the hasty structure of Moslem greatness was tottering to its foundation, when the tumult was seasonably appeased by the disinterested resolution of Omar, who quietly renounced his own pretensions, and offered his hand in token of fealty and obedience to his venerable rival. Perhaps the crafty politician saw in the advanced age of the new caliph but a narrow barrier between himself and the throne.

The Hashemites alone declined the oath of fidelity; and it is remarkable that their chief, Ali, the cousin of their Prophet, and the husband of his favourite daughter, had not, in that numerous conclave, a single voice to advocate his claims. He had publicly refused his concurrence in the inauguration of Abu Beker; but the intrigues of the disaffected could not prevail with him to disturb the peace of his country; nor could the arrogance of Omar, who threatened to consume his habitation with fire, terrify him into submission. The new sovereign despised the pompous epithets of royalty; and, in reverence for the founder of their religion, adopted the simple and more flattering title of caliph. But scarcely had his accession taken place, when he was overwhelmed with the accumulated intelligence of apostacy, revolt, and imposture, from all quarters. The religious spirit which the example of Mohammed had kindled among the Arabs was but a blaze of fanaticism; and on his decease they seemed desirous of returning to the altars of idolatry. The Christians were tributaries rather than subjects; and the obstinate Jews had again directed their hopes to the exploded ritual of Moses. The tribes of the Desert were sinking back into their ancient paganism; and the Koreish would perhaps have restored the images of the Kaaba, had not Abu Beker checked their inconsistency by an effective reproof: "Ye men of Mecca, will ye be the last to embrace, and the first to abandon the religion of Islam?"

The success of the Prophet had encouraged others, even in his lifetime, to emulate his pretensions. Aswad, a dexterous usurper, who seized the government of Yemen, had already suffered the punishment of his rebellion. Moseilama had made alarming progress in the territory of Yemama, and was still in the zenith of his apostacy. A Christian woman, who aspired to the honours of a prophetess (for the spirit of imposture was not confined to the male sex), became his associate; and even in the public camp the decencies of words and actions were spurned by these privileged and amorous saints. Khaled, the scourge of the infidels, with an army of 40,000 men, defeated and dispersed his followers at the battle of Akraba, where 10,000 of the rebels fell; a black slave pierced the upstart apostle with a javelin. His paramour, whose name was Sejaje, returned to idolatry; but afterward became a Mussulman, and died at Bussora. Of the rest of the impostors, some on recantation were admitted to pardon, while others fell victims to their own credulity. By this prompt display of military force the spirit of insurrection was put down. The loyalty of the faithful was revived and confirmed. The wavering tribes returned with humble contrition to the duties of prayer, fasting, and alms; and the religion of the Koran was again believed, and more steadfastly professed, by the whole nations of Arabia. Order and security were restored; but it became necessary to provide an immediate exercise for the restless spirit of the Saracens. On the reduction of Yemama, Khaled marched into Irak and the provinces on the Lower Tigris, where the dominion of the Koran was further extended by a series of rapid and splendid victories. An annual tribute of 70,000 pieces of gold was the first-fruits of these remote conquests.*

From the banks of the Euphrates the general was suddenly recalled to take the command in another quarter. The dying Prophet had meditated the subjugation of Syria; and Abu Beker was only prevented from following up his intentions by the revolt of his own subjects. Events favoured a renewal of the enterprise; and no proposition could have come more welcome to the faithful, burning alike for pillage and the propagation of truth.

The resolution of the caliph was speedily made

* The following are our authorities for the history of the early caliphs and the wars of the Saracens:—Abulfeda (Annal. Moslem. a Reiske, 4 vols. 4to); Elmacin (Hist. Saracen. ab Erpenio); Abulfarage (Hist. Compend. Dynast. a Pococke); Eutychius (Annal. a Pococke, 2 vols. 4to); D'Herbelot (Biblioth. Orient.); Mod. Univ. Hist. vols. i. ii. iii.; Ockley (Hist. of the Saracens, 2 vols. 8vo), who, to the shame of English patronage, died in Cambridge jail instead of Cambridge University. Marigny (Hist. des Arab. 4 vols.); Price (Retrospect of Mohammedan Hist. 4 vols. 4to), whose work is a copious and valuable mine of original authorities. The Greek writers on that period, Theophanes, Zonaras, Cedrenus, &c., may be consulted in Niebuhr's Collection of the Byzantine Historians

known. His circular to the tribes kindled the flame of pious and martial ardour in every province: "This is to acquaint you that I intend to send the true believers into Syria, to take it out of the hands of the infidels; and I would have you know that fighting for religion is an act of obedience to God." From Mecca, and the distant shores of Yemen and Hadramaut, bands of intrepid volunteers hurried to the capital; complaining less of the insupportable heat, and the scarcity of provisions, than the inaction of their master.—When their ranks were complete, he gave his particular instructions to Yezid, son of Abu Sofian, whom he had appointed general of the forces:—To avoid injustice and oppression; to study to deserve the love and confidence of the troops; to acquit themselves like men fighting the battles of the Lord; to spare fruit-trees, cattle, and corn-fields; to stand to their engagements, and never to stain their victory with the blood of women and children. "As you go on," he continued, "you will find some religious persons that live retired in monasteries, who propose to themselves to serve God that way. Let them alone, and neither kill them nor destroy their buildings. And you will find another sort of people that belong to the synagogue of Satan, who have shaven crowns; be sure you cleave their sculls, and give them no quarter till they either turn Mohammedans or pay tribute." This exemption in favour of the monks is by some alleged to have been in return for their hospitality to the Prophet in his youth, or their assistance in compiling the Koran; the terrible doom of the secular clergy was only fulfilling a precept of their religion.

The news of these preparations reached Heraclius; but it was in vain that he appealed to the pride and courage of his subjects, and represented to them the shame of allowing a warlike nation to be insulted by the contemptible Arabs. In the first skirmish the Christians lost their general, with 1200 men. The messengers from Yezid daily announced new successes, and a present of the spoil was despatched to the caliph as the first-fruits of their expedition. The zeal or the avarice of the Arabian chiefs was attracted by the prosperity of their countrymen. A fresh reinforcement was levied for the subjugation of Palestine, the command of which, after some dispute, was delegated to Amru. Zaid was disappointed of the commission; but fanaticism overruled all regard for personal interest. Nowhere was this spirit more remarkable than among the early Saracen generals. Had not the propagation of the Koran exalted them above all private considerations, or had they been actuated by that rivalship and animosity which divided the successors of Alexander, the power of the caliphs must have fallen back to its original insignificance, and the legions of the Faith, instead of marching onward to empire, might have met an inglorious defeat from the swords of contending factions. Abu Obeidah had superseded Yezid; but he was less fortunate than his predecessor, and a single reverse condemned him as unfit to hold the supreme command.

In all emergencies of war, the superior genius of Khaled marked him out for the post of honour; and whoever might be the choice of the prince, he was, both in fact and fame, the foremost leader of the Saracens. From Irak he was recalled to Syria, and his arrival altered the aspect of the campaign. Aracca, Tadmor, and Hauran had submitted; but a rash attempt on Bosra had nearly turned the tide of conquest. Serjabil had been despatched by Obeidah with a body of 4000 cavalry, and was repulsed from its gates with loss. Khaled, with 1500 horse, came in time to rally the flying detachment, and saved the believers from the disgrace of a total overthrow. "See, the villains come!" he exclaimed, as the opened gates of the fortress poured forth their martial array into the plain; "they know we are weary; but let us go on and the blessing of God go with us." The battle grew hot; shouts of Allah akbar! Alhamlah, alhamlah! Aljannah, aljannah! (God is great! Fight, fight! Paradise, paradise!) inflamed the enthusiasm of the Saracens, and threw the ranks of the Christians into disorder. With the loss of 230 men, the Arabs remained masters of the field, and the remnant of the enemy fled to their fortress. The capture of the city was accelerated by the duplicity of Romanus the governor, who, to secure his wealth, or save his life, turned both traitor and Mussulman. The conquest of Bosra opened a way for the siege of Damascus. The Grecian emperor was at this time at Antioch, and, alarmed at the devastations of the Saracens, he sent 5000 men to the defence of the place.

Amid the groves and fountains of this celebrated capital "the hungry Arabs" for the first time pitched their camp. In rude and superstitious ages, a hostile defiance was frequently sent and accepted by the generals themselves, as an omen of future fortune. The personal valour of Khaled was twice displayed in single combat before the two armies, and both the Christian leaders in succession, Azrael and Calous, became his prisoners. Several actions more or less sanguinary were fought; but the Damascenes, finding that every sally only exposed them to fresh losses, resolved to reduce their city to a closer defence. Six weeks of distress compelled them to offer the Moslem leader 1000 ounces of gold and 200 silken robes, on condition of his raising the siege; but Khaled was inexorable, and would listen to no terms but annual tribute or the Koran. They had contrived to drop a messenger from the ramparts, who found his way to Antioch, whence the emperor despatched to their relief 100,000 men, with Werdan at their head. The tumultuous joy of the besieged at the prospect of this seasonable succour revealed the secret to the vigilant enemy; the reinforcement was intercepted, and defeated by a detachment under

Derar, who took with his own hand the standard of the Christians, bearing the sign of the cross, and richly adorned with precious stones.

A new levy of 70,000 men, under Werdan, was again assembled in the neighbourhood. These formidable preparations required the junction of the Saracens, who were dispersed on the frontiers of Syria and Palestine. Yezid was at Belka, Nooman at Tadmor, and Amru had not left Irak. In a circular to the latter, Khaled urged him to join his brethren without delay in their march against the infidel Greeks, "who proposed to extinguish the light of God with their mouths." The other generals had similar orders; and by a singular coincidence in time, the whole forces met on the plain of Aiznadin on the same day (A. H. 11, or the 13th of July, 633); a circumstance which they piously ascribed to the blessing of Providence. On retiring from Damascus, Khaled led the van, in compliance with the wishes of Obeidah, though he offered to take the more perilous station of the rear-guard. The event proved the disinterestedness of his bravery, and the soundness of his judgment; for the besieged, seeing the enemy depart, ventured out with 6000 horse and 10,000 foot, and overtaking Obeidah, defeated him, and carried off a great part of the baggage, with the women and children. The presence of Khaled put a stop to the rout, and of the Christian army only 100 horsemen returned to Damascus. The captives were retaken; among them was the beautiful Khaulah, sister of Derar, and a troop of heroines from Yemen, who were accustomed to ride on horseback and fight like the Amazons of old. The intrepidity of Khaulah saved the honour of the martial sisterhood. They were disarmed of their swords and bows, but with their tent-poles they kept their infidel ravishers at bay till relieved by their own friends.

Forty-five thousand Moslems mustered on the

field of Aiznadin. The troops of Werdan consisted chiefly of cavalry, and have, by historians, been called indifferently—Syrians, from the place of their birth,—Greeks, from the religion and language of their emperor,—and Romans, from the proud appelation still claimed by the successors of Constantine The armies were in sight of each other, and encouraged to their duty by their respective generals Khaled put it to the conscience of his Moslems "to fight in good earnest for religion, or turn their backs and be damned." To the squadron of ladies, Khaulah, Opheirah, and others whose names the annalists have ungallantly omitted to record, he addressed himself,—"Noble girls! be assured that what you do is very acceptable to God and his apostle. You will hereby purchase a lasting memorial, and the gates of paradise will be open to you."

Werdan represented to his troops the shame of slavery, and exhorted them to implore Heaven for succour. To his sudden surprise he was approached by a fierce and half-naked warrior, whom he imagined to be a spy, and sent a party of thirty horse to seize him. It was the intrepid and adventurous Derar, who had undertaken to view the state of the enemy. In his retreat he maintained a successful skirmish against the whole party, and, after killing or unhorsing seventeen of their number, reached his companions in safety. On the eve of battle, a grave elder from the Christian army offered to purchase peace by the gift, to each Saracen soldier, of a turban, a robe, and a piece of gold; ten robes and one hundred pieces to their leader; one hundred robes and one thousand pieces to the caliph. "Ye Christian dogs!" replied the indignant Khaled, "you know your option—the Koran, the tribute, or the sword. As for your proffer of vests, turbans, and money, we shall in a short time be masters of them all."

The onset on both sides was sustained with vigour,

and the battle, or rather the slaughter, continued till evening. Fifty thousand of the Christians fell under the scimitars of the enemy, and the remains of the imperial troops fled—some to Cæsarea, others to Damascus or Antioch. Four hundred and seventy martyrs were left on the field. The spoil was inestimable, including banners, crosses of gold and silver, chains, precious stones, suits of armour, and rich apparel. The glorious intelligence was immediately transmitted to the caliph, and diffused universal joy. The Meccans and other tribes, hostile to the first preaching of the Koran, were now eager to thrust their sickles into the bloody but lucrative harvest of Syria.

The tidings of this defeat filled the Damascenes with grief and terror; and from their ramparts they beheld the ferocious conquerors return to the siege. Amru led the van at the head of nine thousand horse; and the rear was brought up by Khaled in person, with the standard of the black eagle. The wretched citizens had made every preparation during this short respite. For a time their courage was revived by the example and authority of Thomas, an excellent soldier, though living in a private station, and son-in-law to the Grecian emperor. He affected to despise the "contemptible Arabs—poor wretches, going about with hungry bellies, naked and barefoot"—and advised the citizens to attempt a sally next morning, and defend themselves to the last, rather than surrender. Watch was kept during the night, and the absence of the sun was supplied with numberless lights placed in the turrets. At the gate through which they were to march a lofty crucifix was erected in sight of both armies, and before it the bishop with his clergy placed a copy of the New Testament, on the cover of which, as he passed, Thomas imprecated the overthrow of the oppressors, and prayed the Author of that divine book to defend his servants and vindicate his truth.

The onset of the Greeks was met with firmness by the Saracens, whose charges were tremendous and irresistible. The brave Aban fell by a poisoned arrow from the hand of Thomas, an unerring archer; but his death was revenged by his wife, a heroine who could handle the spear and the bow with equal dexterity. In the hottest of the battle she sought the place where his murderer fought; her first arrow pierced the hand of his standard-bearer, though the engines from the walls poured stones and missiles on the besiegers thick as hail; her second wounded Thomas through the eye while engaged in single combat with an Arab who had seized his fallen standard. The wound was dressed on the rampart, and the undaunted champion of the Christians refused to quit the field till night separated the combatants. The devoted widow washed the corpse of her husband, and without a groan or a tear buried him with the usual rites:—"Happy, happy art thou, my dear! thou art gone to the Lord, who first joined us, and hath parted us asunder. Henceforth shall no man ever touch me more; for I have dedicated myself to the service of God."

The citizens were disheartened; their patience and their provisions began to be exhausted; and after a siege of seventy days, the bravest of their chiefs yielded to the hard dictates of necessity. Khaled was inexorable, and declined the chance of capitulation, lest the "Christian dogs" might stipulate for their lives and fortunes, and his soldiers be deprived of their plunder. In the mild virtues of Obeidah the besieged had some hope. At midnight a deputation of one hundred of the clergy and principal inhabitants were introduced to his tent, where they were courteously received, and obtained a written agreement that hostilities should cease—that such as chose might depart with as much of their effects as they could carry away—and that, on paying tribute, the rest should enjoy their lands and houses, with the use and possession of seven churches. On these conditions he was allowed to enter the town by the gate nearest his camp, where the necessary hostages were delivered into his hand.

Of these transactions Khaled was entirely ignorant; and, at the time the truce was concluded, he was storming the walls on the opposite side. By the treachery of a priest, who pretended to have discovered in the book of Daniel the future greatness of the Saracen empire, a party of one hundred Hamyarites were secretly conveyed into the town, and by their means the remainder of the army effected their entrance. The horrid tecbir (the Arabian war-cry) of Allah akbar announced to the astonished Christians that their city was lost. The weapons dropped from their hands as they heard the cry of "No quarter!" from the ferocious Khaled. The ruthless scimitar fleshed itself to the full, and a torrent of Christian blood poured down the streets of Damascus.

The slaughter continued till they reached the church of St. Mary, where the sanguinary conqueror beheld with indignation and surprise the peaceful Obeidah and his troops, with their swords in their scabbards, and surrounded by a multitude of priests and monks. An angry remonstrance ensued between the two generals; the one urging his articles of treaty and the faith of Mussulmans—the latter threatening, in right of his office as general, to put every unbeliever to the sword. The rapacious and cruel Arabs would have obeyed the welcome command, but Obeidah averted the atrocious massacre by a decent and dignified firmness. A council of war assembled in one of the churches; when it was agreed, after violent altercation, that the part of the city which had surrendered to Obeidah should be entitled to the benefit of his capitulation; and to this pacific measure Khaled reluctantly assented until a final decision should be pronounced by the caliph.

A large majority of the people accepted the proffer of toleration and tribute, and remained in their ancient habitations.* But Thomas and the valiant patriots who fought under his banner preferred the wretched alternative of exile. A protection of three days was granted them, but to extend only to the country in possession of the Moslems. Khaled endeavoured to exclude the miserable refugees from the full benefit of the treaty, by limiting their exports simply to provisions; and sternly declared, that at the expiry of the three days they might be pursued and treated as enemies. In a meadow in the suburbs a large encampment was formed, where in haste and terror the exiles collected their most precious moveables in plate, jewels, and apparel; including the imperial wardrobe, in which there were above three hundred loads of died silks.

The fall of Damascus was communicated to the caliph, but he lived not to receive the joyous intelligence; he died on the very day the city was taken (Friday, the 3d of August, A. D. 634), after a short reign of two years and three months. The manner of his life was simple, austere, and frugal. When he assumed the pontificate, he ordered his daughter Ayesha to take a strict account of his private patrimony, that it might be seen whether he had been enriched or impoverished by the service of the state. All he claimed for himself was a stipend of three drachms or pieces of gold, with sufficient maintenance for a camel and a black slave. The surplus of his exchequer, as well as of the public money, was every Friday distributed, first to the soldiers and the most deserving, and next to the most indigent.

The disturbances which attended his own accession, Abu Beker happily prevented by a testamentary appointment of Omar. The latter at first seemed willing, like the reluctant Cæsar, to refuse the kingly crown, as having "no occasion for the place."—"But," replied the other, "the place has need of you." The title of Caliph of the Caliph of the Apostle of God being deemed tautological, and likely to increase to an inconvenient length, was exchanged for that of *Emir el Mumenin*, or Emperor of the Faithful. His first measures were to follow up the military operations of his predecessor; and after a short expedition into Irak, with indifferent success, Obeidah continued to prosecute the war in Syria.

The Damascene exiles had been four days and nights on their march, and might have retired unmolested, but for the imprudence or passion of a youthful lover. A noble citizen, named Jonas, was betrothed to a wealthy maiden—but her parents, on some slight pretext, delayed the consummation of their nuptials. During the pressure of the siege the danger of their situation induced them to attempt their escape. Having bribed the nightly watch at the gate Keisan, Jonas rode forward to lead the way, but was surrounded and seized by a squadron of the patrol under Derar. Another horseman followed (it was the lady); but the signal in Greek, which the Arabs did not understand,—"The bird is taken!" admonished his adventurous mistress to hasten her return. He was brought before Khaled, who promised him his wife, when the city was taken, on condition of his embracing the religion of the Koran; if not, he must die on the spot. The lover chose apostacy—professed his belief in Mohammed, and continued in the enemy's camp to perform the duties of a brave and sincere Mussulman. The lady, whom the tragic muse has named Eudocia, in despair and detestation of her apostate lover, shut herself up in a nunnery.*

* On the fate of these lovers, whom he calls Phocyas and Eudocia, Mr. Hughes has constructed his tragedy the Siege of Damascus.

When the city surrendered, Jonas flew to the monastery where Eudocia had taken refuge; but his name was forgotten, his passion scorned: she preferred her religion to her lover, and took her departure in the caravan of fugitives; bidding adieu to a country which was no longer free, and an attachment which could only be remembered with sorrow. Jonas was inconsolable; and, in the hope of rescuing his wife, urged the conqueror to pursue the weary exiles who, he assured him, might yet be overtaken, and offered his services as guide. Khaled yielded to his importunities; and, at the head of four thousand horse in the disguise of Christians, took the route of the fugitives. In a wet and dark night they traversed the rocky passes of Mount Lebanon; and at the dawn of day they beheld, in a pleasant valley below, the tents of Damascus shining in the morning sun. Of the hapless wanderers, some lay overcome with fatigue and sleep, others were spreading their drenched garments to dry. Khaled divided his troop into four squadrons, who rushed successively on the unarmed and promiscuous multitude. The Christians defended themselves with bravery; but the loss of their general, whose head was insultingly mounted on the standard of the cross, announced that victory had declared for the enemy. One man only escaped the fury of the barbarians, to carry to the court of Constantinople the dismal story of the catastrophe.

In the tumult of the battle, Jonas sought and found the object of his fatal pursuit. The indignant lady repulsed his advances sword in hand. After a hard-fought combat she became his prisoner; but, loathing his perfidious embraces, in a fit of tranquil despair she struck a poniard to her heart. The widowed daughter of Heraclius, a princess of admirable beauty, was taken captive, and offered to the disconsolate lover; but he yielded to the wishes of Khaled, who spared and released her without a ransom; sending her away with a haughty message of defiance to the emperor, that he should never enjoy peace till his daughter and his dominions were in possession of the Saracens. The bereaved Jonas fought and fell in the service of his adopted religion. To encourage proselytes, Raphi, a brother officer, declared he had seen the blessed martyr in a vision walking in the verdant meadows of paradise, with gold slippers on his feet, and forgetting the love of Eudocia in the embraces of seventy virgins, so bright and fair that, had one of them but half unveiled her charms in this world, the sun and moon must have paled before the splendour of her beauty.

Abu Obeidah had displaced Khaled in the command; but he had no fixed plan of operation, and his wavering policy was uncertain whether to direct the march of the believers to Jerusalem or Antioch. An expedition was suggested and undertaken, to which the Arabs were solely prompted by their insatiable cupidity. The fair of Abyla, or the Monastery of the Holy Father, a place near the eastern base of the Anti-Libanus, about thirty miles from Damascus, was annually celebrated at Easter by a vast concourse of merchants. The cell of a devout hermit attracted crowds of pilgrims; young and old, rich and poor, solicited his blessing; and no married couple thought their conjugal felicity complete till they had received his benediction. In the hope of an easy conquest and a large spoil, Abdallah was despatched with a handful of five hundred cavalry.

This festival happened that year to be honoured by the nuptials of the daughter of the governor of Tripoli; and the usual crowd was swelled by an escort of 5000 horse that attended the person of the bride, who had come to have her faith and her fertility confirmed by the pious anchorite. The invaders were awe-struck, but shame prevented their retreat; and their drooping courage revived in the morning when the fair commenced, and the tempting merchandise was spread before their eyes. The reverend prior had begun his sermon, attended by a vast throng, among whom were many of the nobility and officers richly dressed. The avarice of the Saracens was inflamed. "Paradise," exclaimed Abdallah, "is under the shadow of swords: either we shall succeed and have the plunder, or die and gain the crown of martyrdom." The rapidity of their onset gave them the first advantage; but they were encompassed and almost overwhelmed with numbers so soon as the enemy had recovered from their surprise. A hasty message informed Khaled that the believers were in danger of being lost. With a troop of cavalry he flew to their relief; and about the hour of sunset a cloud of dust announced their approach to the weary combatants, whose diminutive band, amid the swarms of their assailants, is fancifully likened to a white spot on a black camel's skin.

Their arrival changed the fortune of the day; shouts of Allah akbar rent the skies; the Christians were dispersed and pursued with great slaughter as far as the river of Tripoli, whose waters saved them from the vengeance of the conquerors. The various merchandise of the fair, the fruits and provisions, and the money brought to purchase them, became the prey of the spoilers. The monastery, in which the governor's daughter and forty of her waiting-maids were taken, was profaned and robbed of its plate, curtains, and gay decorations; among which, we are informed, was a cloth of curious workmanship, embroidered with an image of the blessed Saviour, which was transported to Yemen, and sold for ten times its weight in gold. Horses, asses, and mules were diligently loaded with the wealthy plunder, and the glutted barbarians returned in triumph to Damascus.

The luxuries of Syria had begun to corrupt the

abstemious Arabs, and the messenger that carried to Omar the news of this second victory, reported that the Mussulmans had learned to drink wine. A punishment of fourscore stripes on the soles of the feet was ordered to be inflicted on the offenders; and so tender was the conscience of the believers, that on the proclamation of Obeidah, numbers submitted without an accuser to the penance of the law

Terror had already spread the fame of the Saracens beyond their actual conquests; though in the prosecution of the war their policy was not less effectual than their swords. The cities of Syria individually trembled for their security. Instead of acting in concert, each was willing to make the fall of others the signal for their own capitulation, and agreed to purchase a temporary respite at an enormous ransom, which only enriched the enemy by impoverishing themselves. Chalcis alone was taxed at 5000 ounces of gold, as many of silver, 2000 robes of silk, and as many figs and olives as would load 5000 asses. The less wealthy or less obstinate paid in proportion. By these short and separate treaties the union of the Christians was dissolved; their hands were tied up from mutual assistance while the Arabs were ravaging the country; and at the expiry of the truce their exhausted magazines and arsenals left them an easy prey to the besiegers.

Homs or Emesa, and Baalbec or Heliopolis, both populous and wealthy cities, were the next that yielded to the rapacity of the barbarians. But the slowness of their progress was offensive to the caliph, who wondered at the silence and inactivity of his soldiers. In an epistle to Obeidah he gently insinuated his suspicions that the wives and the spoil of Syria were dearer to them than the service of God and his apostle. The Moslems understood the rebuke, and with tears of rage and remorse demanded to be led forth to the "battles of the Lord."

Repeated messages of defeat and disaster informed Heraclius, then at Antioch, of the success of the insolent barbarians. To arrest their progress, and drive the robbers of the desert for ever from his dominions, an army of 80,000 men was collected from the provinces of Europe and Asia. Cæsarea, Tyre, Sidon, Joppa, Tripoli, and other coast-towns were strongly garrisoned. The main body was intrusted to Manuel, one of the bravest officers in the service. He was reinforced by 20,000 Christian Arabs, with Jabalah, king of Gassan, at their head. This prince had embraced Islam in the presence of Mohammed; but afterward, having quarrelled with a person in the Temple at Mecca, he had abandoned the Koran, fled to Heraclius, and, in a letter from Omar to the Syrian army, he was publicly denounced an apostate. The banner of the Gassanites was planted in the van; for it was a maxim of the Greek general "to cut diamond with diamond;" in other words, to oppose the fury of the barbarians with the valour of their own countrymen.

Manuel took immediate possession of Emesa, prematurely evacuated by the Moslems, and advanced to the banks of the Yermouk, where Khaled had taken his position, and where he resolved to stake the fate of Syria. This petty stream (the Hieromax of the Greeks), immortalized by one of the most sanguinary battles of antiquity, rises in Mount Hermon, and winding through the plain of Decapolis, is lost, after a short run, in the Lake of Tiberias. The tardy policy of the Grecian general lost him the only chance he ever had of driving the invaders back to their deserts. Instead of attacking them before the arrival of 8000 auxiliaries from Medina, he wasted the time in useless overtures for peace, and allowed the enemy to gain some advantage by routing one of his detachments on their march. An order from the emperor, proposing to try the effect of a conference, was communicated to Obeidah, and Khaled was instructed to repair to the Christian camp. Addressing himself to the rapacity of the Moslems, Manuel engaged to secure the payment of a sum of money, provided they would withdraw from the country: to Omar, as their sovereign, 10,000 dinars* (4625*l.*); to Obeidah, one-half that sum; to a hundred of his principal officers, 1000 dinars each; to every horseman, 100; and to every foot-soldier, 50. Khaled, on his part, offered to their choice the usual conditions of conversion, tribute or the sword; recommending, as the best means of averting further calamity, that "they should admit the lamp of eternal truth into their habitations;" but the offer was rejected with scorn.

Both sides were prepared for action. Manuel disposed of his multitudinous legions into several divisions, each consisting of 20,000 men. Mounted on a black charger, and clad in gorgeous armour, he conducted the movements of the centre in person. The van of the Moslems was led on by Khaled, restored once more to the supreme command; while his colleague, with the yellow flag under which Mohammed had fought at Khaibar, was posted in the rear, that the flight of the believers might be checked by the presence of this consecrated ensign. Khaulah and her band of Amazons were stationed in the same line, and for the same reason. The exhortation of Khaled was of tremendous brevity: "Paradise is before you; the devil and hell-fire behind!" Their march to the field was in profound and awful silence; but the ranks of the Christians presented a different scene. The mingled noise occasioned by

* The gold dinar has been differently estimated, from 5*s.* to 13*s.* 6*d.*—we have taken a medium, or 9*s.* 3*d.*. The drachm of gold is reckoned equivalent to about 5½*d.* In speaking of the plunder of the Saracens, we may remark, once for all, that according to Major Price, the value of money then may be taken as ten times greater than at the present day.

the priests chanting their gospels, and the motions of their beads and chaplets, is compared to the roll of distant thunder; while the glittering of their armour resembled flashes of lightning.

Their first effort was directed against the centre of the Arabs, which by a violent assault they endeavoured to penetrate; but the whole of the division was annihilated by Khaled with a body of 10,000 horse. Another and an equally unsuccessful attempt was made; yet such was the weight of the Roman cavalry, that the right wing of the Saracens was broken and separated from the main body. Thrice did they retreat in confusion, and thrice were they driven back to the charge by the blows and reproaches of the women, whose tent-poles were more effective than the terrors of everlasting punishment. The battle raged for several days; and night separated the combatants only to renew the encounter. In the intervals of action many a lance was shivered in single prowess; and such was the cunning and skill of the Armenian archers, that in one day, the Day of Blinding, 700 of the Moslems lost one or both of their eyes,—a deformity which, instead of considering a misfortune, they gloried in as a mark of Divine favour.

The carnage at Yermouk surpassed that of any preceding battle: the veterans of former wars confessed it to be one of the most desperate and doubtful encounters they had ever seen; but it was decisive. Seventy thousand of the best troops of Heraclius, with their general, and a vast number of his principal officers, were left on the field. Of the fugitives who attempted to escape many perished during the confusion of the night in the woods and ravines, or in the precipitous watercourses that intersect the adjoining mountains. Besides those that were slaughtered after the defeat, many found a watery grave in the Yermouk, which in their panic they had endeavoured to cross. Altogether, the

Christians are stated to have lost 150,000 men in killed, and 48,000 in prisoners. Perhaps the Arabs may exaggerate in point of number; but the Greek historians themselves have admitted the total overthrow of the Roman power in Syria in this sanguinary engagement (Nov. 636), which they long after bewailed as a just retribution for their sins.

Of the Saracens, 4030 martyrs were buried on the spot. Next to the prayers of the caliph, the glory of the triumph has been ascribed to the female warriors; for their ablest generals confessed that their bravery had been in vain, had not the battle been again and again restored by the firmness of the women. The intelligence of the victory, together with the legal proportion of the spoil, were speedily conveyed to the throne of Omar, and received with the most lively demonstrations of joy and gratitude. After detailing the loss on both sides, "I found," says Obeidah in his letter, "some heads cut off; not knowing whether they belonged to Mussulmans or Christians, I prayed over them, and buried them. The numbers drowned are unknown to any but God; as for those that fled to the deserts and mountains, we have destroyed them all."

Recovered from the toils of the campaign at Damascus, the Saracens were eager to be led to new conquests. The dispersion of the Christians left them free to choose which of the fortified towns should be the object of their first attack; and, in obedience to the caliph's commands, they proceeded forthwith to the reduction of Ælia* or Jerusalem. Trusting to their engines and the sanctity of the place, the fearless inhabitants disdained to reply to the pacific messages of the enemy, and prepared for a vigorous defence. Ten days were wasted in prayer and vain expectation; and, on the eleventh, the town was beleaguered by the whole strength of the

* A name derived from the colony of the Emperor Ælius Hadrian.

Saracen army. The citizens maintained an obstinate siege of four months,—not a single day of which passed without fighting. The patriarch Sophronius imprecated from the walls the Divine displeasure on the disturbers of the Holy City. The warlike machines showered their missiles incessantly from the ramparts; while the difficulties of the besiegers were increased, and their ranks thinned by the inclemency of winter.

The perseverance of the enemy at length induced Sophronius to demand a conference, in the hope of obtaining the terms of an honourable capitulation. Obeidah was equally ready to treat; and the inhabitants consented to surrender on the singular condition of receiving the articles of their security and protection from the hands of the caliph, and not by proxy. This strange proposition was communicated to Omar; and after some discussion, he resolved immediately to visit the ancient capital of Palestine. The rude simplicity of his equipage and manners is minutely described by the Moslem writers, and presents a striking contrast to the gaudy pageantries that usually surround the haughty monarchs of the East. On this occasion the emperor of the faithful, the conqueror of Syria, Persia, and Egypt, courted no distinction beyond the meanest of his subjects. His dress was a coarse woollen garment, with a scimitar hung from one shoulder, and a bow on the other. He rode a red camel, carrying with him a couple of sacks, one filled with fruits, the other with sodden barley in the husk,—a sort of provision called *sawik*, and in general use among the Arabs. Before him he had a leathern bottle of water, and behind him was suspended a large wooden platter. When he halted on the way, the company was uniformly invited to partake of his homely fare; and the humblest of his retinue dipped their fingers in the same dish with the mighty successor of the Prophet. The spot where he reposed for the night was never abandoned in the morning without the regular performance of prayers. On the journey he redressed the wrongs of some poor tributaries, rebuked the licentious polygamy of the ignorant Arabs, and chastised the luxury of the Saracens, by stripping them of their rich silks, the spoils of Yermouk, and ordering them to be dragged with their faces in the dirt. At Joppa, five stages from Jerusalem, he was met by Obeidah, with the principal officers of the army. Apprehensive lest the meanness of his appearance,—for according to some writers he was leading the camel by the bridle, while the slave was mounted in his turn, the animal being their joint property,—should excite contempt in the eyes of a nation long habituated to more polished observances, they prevailed with him to attire himself in a suit of white apparel, and accept the horse which they had provided for introducing him among his newly-conquered subjects. But he soon felt the encumbrance of this novel equipment; and, resuming his barbarous guise, he entered the camp before Jerusalem, exclaiming against the absurdity of forsaking established usages for the mere gratification of a vain and ridiculous caprice.

The articles of capitulation, by virtue of which the inhabitants were entitled to the free exercise of their religion, to their properties, and the protection of the caliph, were signed; and the sovereignty of the place and of the whole adjacent territory was vested in the conqueror. In these conditions, the basis of most of the treaties since granted by the Mohammedan princes to the Christians, a broad line of distinction is drawn between the followers of the Cross and of the Koran. The former were to admit the latter into their churches at all times: to entertain them gratuitously on their journey for the space of three days: to rise up, as a mark of respect, when they are disposed to sit; to avoid the same dress, names, language, and forms of salutation; to renounce the use of saddles, arms, and intoxicating liquors; to have no crosses in their churches, streets, or books; and not to ring, but merely to toll, their bells.

On the ratification of the terms, the gates were thrown open, and the Moslems, with Omar and the officers of the army at their head, took possession of the city. The Christian patriarch entered by the side of the caliph, who conversed with him familiarly, putting many questions concerning the religious antiquities of the place. They visited the Church of the Resurrection together; and, at the hour of prayer, the caliph declined offering his adorations in the temple, preferring the steps of the porch; where he spread his mat, and performed his devotions.

Thus fell Jerusalem, the glory of the East, the imperial seat of David and Solomon. The number of the slain and the calamities of the besieged were greater than when taken and sacked by the legions of Titus; yet the servitude of the Romans, either in its condition or its duration, was nothing in comparison with the tyrannical sway of the Saracens, in whose hands it has continued to the present day, except for an interval of ninety years, when the valour of the Crusaders restored it to the possession of the Christians. By command of the caliph, the ground on which stood the Temple of Solomon was cleared of its rubbish, and prepared for the foundation of a mosque, which still bears the name of Omar.

Aleppo, the Beræa of the Greeks, had not then attained the commercial celebrity of later times; but it was a place of considerable strength. The castle was built on a high artificial mound, at a little distance from the town. Obeidah found the inhabitants more disposed for capitulation than defence; and, in the absence of the governor, thirty of the chief merchants secured their lives and religion at a moderate composition. The surrender of the town did not include that of the citadel, which the martial genius of Youkinna, who, at the head of 12,000 men, had routed a detachment of Arabs, was determined to maintain to the last. For more than four months he set the arms of Obeidah and the valour of Khaled at defiance. Three hundred Christian captives were beheaded before the castle-wall; but this execution had no effect in subduing the resolution of the besieged. Omar was applied to for advice, and a reinforcement of fresh troops was despatched, with a train of horses and camels to expedite their march. Among these subsidiaries was Dames, a gigantic Arab, of servile birth, but of invincible strength and courage; and his single intrepidity effected what the united perseverance of the Moslems had failed to accomplish. Weary at seeing their hopes and their patience consumed at the foot of this impregnable fortress, on the forty-seventh day after his arrival he proposed, with only thirty men, to take the place by stratagem. His design was concealed under the appearance of a retreat; and the tents of the Arabs were removed to the distance of three miles from Aleppo. A Greek belonging to the garrison was seized, and from him some useful communications were extracted.* At the dead of night the adventurers crawled to the foot of the wall, where they lay in ambush. Their leader was covered with a shaggy skin, and provided with a hard crust of bread, that, in case of discovery, he might elude suspicion by imitating the noise of a dog gnawing a bone. Having scaled the most accessible part of the precipice, he ordered seven of his companions to mount on each other's shoulders, himself sustaining and raising the weight of the whole column, till the highest reached the top of the battlements. The watchmen, who were asleep,

* "God curse these dogs," said the illiterate slave when he heard the Greek speak, "what a barbarous language they use!"

were hurled to the ground, and despatched by their associates below. Unfolding his turban, the first drew up the second, until they all reached the parapet in safety, ejaculating the while, "Apostle of God, help and deliver us!" With bold and silent step Dames explored the palace of the governor, whom he found in the banquet-room celebrating, in riotous merriment, the pretended retreat of the enemy. Favoured by the darkness of the night, he traversed the castle, stabbed the sentries at every gate, and took possession of their posts without opposition; not, however, without exciting the alarm of the garrison, who surrounded and would soon have cut the intruders to pieces, had they not unbolted the gates, let down the drawbridge, and admitted a detachment of cavalry, which, at daybreak, had advanced to their relief. The Christians threw down their arms, and surrendered at discretion.

The castle, being taken by storm, was pillaged by the Moslems; but, in order to save their wealth, and their wives and children, Youkinna and several of the chief officers became proselytes to the Koran. It was the character of the times to admire acts of desperate and chivalrous enterprise; and on this occasion, the servile condition of Dames was lost in the glory of his exploit. Out of courtesy to their champion, the army did not decamp from Aleppo until he and his fellow-adventurers were perfectly cured of their wounds.

The reduction of Antioch, the seat of the Greek emperors, was the next object of the Moslem conquerors, but this rich capital was still protected by the strong castle of Azaz, garrisoned by 13,000 troops under Theodorus, and the iron bridge of the Orontes. The loss of these defences, and of a partial action which was fought under the walls of the city, decided its fate. The trembling inhabitants purchased their safety by the immediate payment of 300,000 dinars of gold (138,750*l.*). The importunities of his courtiers had for a time retarded the flight of Heraclius; but when he beheld the battlements closely invested by the Saracens, his patriotism yielded to considerations of personal safety. Having assembled the bishops and principal men of the city in the Greek church, he there bewailed the unhappy fate of Syria, which he devoutly ascribed to the sins of the prince and the people. While his ears had been daily assailed with rumours of defeat, his imagination was terrified with dreams of a falling throne, and a crown toppling from his head. Ascending a hill in the neighbourhood, he cast a last look on his beloved Antioch, and the long fruitful valley (the Hollow, or Cœlosyria) stretching away with its flourishing towns and glittering turrets, from north to south, between the snowy chains of the two Lebanons; and with expressions of regret and conviction that he should never more behold these interesting and favourite regions, he made his way with a few domestics to the Mediterranean shore, and privately embarked for Constantinople. A Mohammedan tradition has laboured to make him a convert, by means of a cap sent him by Omar, in which was sewed a text of the Koran; and which cured him of an obstinate headache when every other remedy had failed. History, perhaps with some truth, has recorded a conversation between his imperial majesty and one of the Moslem captives, as to the person and dignity of their sovereign. "What sort of a palace," said Heraclius, "has your caliph?" "Of mud." "And who are his attendants?" "Beggars and poor people." "What tapestry does he sit upon?" "Justice and uprightness." "And what is his throne?" "Abstinence and wisdom." "And what is his treasure?" "Trust in God." "And who are his guards?" "The bravest of the Unitarians."

Constantine, the eldest son of Heraclius, was stationed at Cæsarea, the second metropolis of Palestine; but after the flight of his father and the surrender of so many places of strength, he found himself unable to contend against the united forces of the caliph. Amru with a division of the army infested that part of the country, and was prepared to give battle to the Christians; but the prince, dispirited with losses and afraid of falling into the hands of the Saracens, left his government, and taking shipping in a tempestuous night with his family and his wealth, departed for the tranquil shores of the Bosphorus. The Cæsareans, abandoned by their chief and without the means of defence, with one consent surrendered the city to Amru;—having purchased their security by paying 200,000 pieces of gold.

Obeidah, fearing lest the luxuries of Antioch might enervate his troops,—for the Grecian women had begun to seduce the stern virtues of the Arabs,—withdrew his army after a brief refreshment of three days. But Omar was more indulgent than his lieutenant to the infirmities of the faithful. "God," said he, in an epistle mildly censuring him for his unkindness to the Moslems, "hath not forbidden the use of the good things of this life to faithful men, and such as have performed good works; wherefore you ought to have given the Saracens leave to rest themselves, and partake freely of the good things which the country afforded, that whosoever of them had no family in Arabia might marry in Syria, and purchase as many female slaves as they had occasion for." The fall of Damascus, Jerusalem, Aleppo, and Antioch may be said to have completed the conquest of Syria. The mountainous districts of Palestine were overrun by a troop of 300 Arabs and 1000 black slaves, who, in the depth of winter, climbed the snowy ridges of Lebanon. Tripoli and Tyre were betrayed; a fleet of fifty transports, destined for Cyprus and Crete, entering the harbour of

Tripoli, were seized by Youkinna, and yielded to the Saracens a welcome supply of arms and provisions. The towns, or provinces, of Ramla, Acre, Joppa, Ascalon, Gaza, Shechem or Nablous, and Tiberias, surrendered without resistance; and their example was followed by the inhabitants of Sidon, Beirout, Laodicea, Apamea, and Hieropolis. Within six years after their first expedition, and 700 after Pompey had despoiled the last of the Macedonian kings, this fertile and populous region submitted to the rule of the Arabian caliph. Eastward, Khaled had extended his victories, and reduced Beles, Racca, Rahabah, and various other fortified towns on the Euphrates.

The same year that completed the subjugation of Syria visited the conquerors with a dreadful pestilence, more fatal to their ranks than the swords of the Greeks or the luxuries of Antioch. Five-and-twenty thousand of the Moslems, including Obeidah, Yezid, Serjabil, and many of the most distinguished companions of Mohammed, were swept off by the plague of Emmaus (the place where it made its first appearance), which spread its ravages, with a terrible mortality both to men and cattle, over the whole country, as far south as Medina.

Khaled, though he escaped a species of death so unwelcome to a soldier, survived his fellow-conquerors only three years, and ended his days under a cloud of ignominy and injustice. However unrequited by the caliph, the merit of the conquests in Syria and Palestine was by the public voice ascribed to the superior skill and singular prowess of this gallant soldier, the fame of whose exploits had long rendered him the theme of general admiration. One of the poets of the day, who undertook to perpetuate his glory, has celebrated the "terrors of his mace and the lightning of his scimitar, which spread wretchedness and mourning among the cities of the Franks." A charge of embezzlement and of appropriating to himself the public treasure was preferred against him, and too rashly credited by the envious Omar. These suspicions were strengthened by his liberality to the panegyrical verse-maker, and the extravagant dower of 100,000 drachms of gold (2291*l.* 13*s.* 4*d.*) to the beautiful widow of Malec, whom he had married while his hands were yet reeking with the blood of her murdered husband. In the presence of the caliph, he was compelled to reply to the interrogatories of his accusers, with his turban tied round his neck, one end of which was held by the common crier. To this indignity, and a fine of one-half of his effects, he submitted with exemplary moderation; declaring that the dictates of resentment, however just, should not prevail with him to resist the will of his superior. On a second examination, he was condemned to the further payment of 40,000 drachms (916*l.* 13*s.* 4*d.*), being the moiety of all he had been allowed to retain. Such were the multiplied and humiliating mortifications to which this undaunted champion of the Koran was compelled to submit, after rendering so many and important services to the cause of Islam. His last moments were imbittered by the reflection that, after having sought the glory of martyrdom in many a bloody field, and felt the weapons of the enemy in every limb, he should descend to the grave wronged and dishonoured, among the common herd of ordinary mortals. The exhausted state of his coffers constrained Omar to acknowledge that his suspicions had been unjust. The caliph condoled with the aged mother, who was repeating, with tears of anguish, some of the numerous encomiums on her brave but unfortunate son; he visited at Emesa the tomb of the injured conqueror of Syria, and expressed in terms of unavailing regret his esteem for the hero of a hundred battles, whom the hand of death had now placed beyond the reach of envy.

At the demise of Obeidah, the command of the

Moslem army devolved on Amru; and the place of the first conquerors was supplied by a new generation of their children and countrymen. The terrors of the pestilence were lost in the passion for victory and martyrdom which animated the hearts of the Mussulmans. Their eagerness for this sort of reputation may be expressed in the words of an Arabian youth, when tearing himself from the embraces of his mother and sister to join the banner of Obeidah: "It is not," said he, "the delicacies of Syria, or the fading delights of this world, that has prompted me to devote my life in the cause of religion. I seek the favour of God and his apostle; and I have heard that the spirits of the martyrs will be lodged in the crops of green birds, who shall taste the fruits and drink of the rivers of Paradise. Farewell, till we meet again among the groves and fountains which God has provided for his elect." The devastations of the plague rendered the presence of Omar necessary to repair, as far as possible, the desolate state of the northern provinces. During a residence at Ramla in Palestine, he filled the numerous vacancies, regulated the supply of provisions for the soldiers, and assigned to the heirs of the deceased Moslems all estates to which just and equitable claims could be produced. The Greeks had been driven from every part of the extensive tract between the Euphrates and the Mediterranean. To the north of Syria the conquerors passed Mount Taurus, and reduced to their obedience the rich plains of Cilicia, with its capital, Tarsus, the ancient residence of the Assyrian kings. Beyond a second ridge of the same mountains, they spread the flame of war rather than the light of religion, as far as the shores of the Euxine, and the neighbourhood of Constantinople. To the east they achieved the conquest of Diarbeker; thus violating the utmost limit of Augustus, the long-disputed barrier between Persia and Rome. Already had Yezid threatened to cross the Euphrates; but his retreat was purchased by the governor of the province at the enormous annual tribute of 100,000 pieces of gold. Ayaz, at the head of 5000 men, marched into that territory; the walls of Edessa, Amida, Dara, and Nisibis, which had resisted the arms and engines of Shapoor and Nooshirwan, were levelled in the dust; but the victor was recalled, and died at Emesa.

PERSIA, to which we shall next accompany the victorious Saracens, was only saved from an earlier doom by the war in Syria. The wealth of this ancient empire was of itself a sufficient attraction; while its weakness left it an easy prey to the roving bands of "naked lizard-eaters" from the desert. For more than thirty years the reign of Khoosroo Purveez had been marked by a success never surpassed by the most renowned of his ancestors. But his magnificence fell with unexampled rapidity. Within six years he lost all his foreign conquests, and saw his dominions overrun by the legions of Heraclius (A. D. 622), who marched in one direction as far as the Caspian, and in another to Ispahan. The ravages of the Greeks were succeeded by the accumulated evils of famine and anarchy, the disputes of the nobles, and a succession of weak sovereigns; or rather the pageants of rival factions, of whom no fewer than six possessed the throne in the brief space of as many years. In this state of dissension and decay the Moslems found Persia when they first directed their warlike operations towards its frontiers. And we shall perceive in the sequel, that as much time and exertion were expended in achieving the conquest of the narrow slip of country on the banks of the Orontes, as was employed in the subjugation of those opulent and extensive regions which fill the space between the Euphrates and the distant Oxus.

The death of Abu Beker, and the probability of a contested succession, had encouraged the Persian government to attempt a more effectual resistance to the encroachments of the Arabs; and even to expel them from their usurpations in Irak, where Mothanna presided over the interests of the caliph and the Koran. The attention of Omar was called, at the commencement of his reign, to the "golden soil of Chaldea," so famed for its fertility, the magnificence of its cities, the variety of its manufactures, and the multitude of its flocks and herds. The avarice of the believers was stimulated by the illusions of dreams and the exhortations of prayer; and the ancient awe of the power and resources of the Persian monarchs faded away before the dazzling splendour of conquest and spoliation. Amru, Obeid, and Saleit were despatched with fresh supplies from Medina, to join the troops in Irak; and their first victory was over two small detachments of the enemy, commanded by Jaban and Roostum; the latter showing great personal gallantry, by killing with his own hand several of the bravest of the Moslems. But the main army, amounting to 80,000 under the command of Jalanous or Galen, was on its march, and took post on the eastern bank of the Euphrates, opposite the bridge constructed by Obeid, whose rashness proved fatal to the Saracens. Dispirited by the death of their leader, they fled in confusion. Numbers were slain, and 4000 drowned in attempting to recross the river.

Mothanna, who made a good retreat, communicated to the caliph the news of this disaster, which had nearly lost for ever the fruits of Khaled's victories. A fresh levy marched into Irak under Jarir; and in a second encounter near Hira, which lasted from noon till sunset, the enemy were put to flight, Mahran, their general, having fallen in single combat by the hand of the Moslem leader. The pursuit was most sanguinary, and was remembered as the Day of Decimation, every Mussulman being said to have slain ten of his adversaries; so that the slaughter of the enemy may be estimated at 100,000 men. Their want of success the fickle Persians attributed to the incapacity of their queen, Arzemidocht, a daughter of Khoosroo. "This we get," they murmured, "by suffering a woman to rule over us;" and in the hope of bettering their fortune, the throne was immediately transferred to Yezdijird, a descendant of the renowned Nooshirwan. The first measure of the youthful monarch was to send an envoy to Saad, the leader whom the caliph had appointed to the chief command. A deputation of three old Arab chiefs repaired to Madayn, the head-quarters of the Persian army. Their mean appearance excited the disdain of the luxurious monarch; for they wore the striped camlets of their country, had small whips in their hands, and rude sandals on their feet. "We have always," said he, "held you in the lowest estimation. Arabs hitherto have been known in Persia only in two characters; as merchants and as beggars. Your food is green lizards; your drink salt water; your covering, garments made of coarse hair. But of late you have come in numbers to Persia; you have eaten of good food; you have drunk of sweet waters; and have enjoyed the luxury of soft raiment. You have reported these enjoyments to your brethren, and they are flocking to partake of them. You appear to me like the fox in our fable, who went into a garden where he found plenty of grapes. The generous gardener would not disturb him, and thought the produce of his vineyard would be little diminished by a poor hungry fox enjoying himself. But the animal, not content with his good fortune, went and informed all his tribe. The garden was filled with foxes; and its indulgent master was forced to bar the gates, and kill all the intruders to save himself from ruin. However, as I am satisfied that you have been compelled to this conduct by absolute want, I will not only pardon you, but load your camels with wheat and dates, that when you return to your native land you may feast your countrymen. But be assured, if you are insensible to my generosity, and still remain in Persia, you shall not escape my just vengeance."

The Arabian messengers heard unmoved a speech displaying at once the extremes of pride and of weakness. They did not palliate or deny the scanty resources of their country; and briefly offered to his choice the Koran, tribute, or the sword. Yezdijird was too proud to listen to such degrading terms; the embassy was dismissed, and war resumed with all the vigour of which the declining empire was capable. The hosts of Persia were as numerous and as feeble as in the days of Darius; and on this occasion the force of the great king has been estimated at 120,000 men, the command of which was intrusted to Roostum. The Saracens were strengthened by recruits from Arabia and Syria,—their whole army being augmented from 12,000 to 30,000 troops, the best soldiers the East had ever seen.

The plain of Cadesia or Kudseah, lying on the skirts of the Desert, about two stages from Cufa, was the scene of action; but four months were spent by the Persians in negotiating, and devising vain expedients to protract hostilities without the hazard of a battle.

The Arabs were disposed in three lines; Saad having directed his captains to consider the first tecbir which they should hear him utter from his post, a terrace of the castle, as the signal to adjust their ranks; the second, to fix their arrows in the level, to couch their lances, and draw their swords; and the third to rush upon their adversary. Various skirmishes, and trials of individual valour in single combat, took place in the interval between the two armies; but the battle itself lasted for several days; each distinguished by its peculiar appellation. On the first, called the Day of Concussion, darkness put an end to the contest; both sides retiring to their encampments without claiming any advantage. With the morning sun, the conflict was renewed; and while the armies were engaged with equal fury and obstinacy, the crimson banners of the advanced guard of the Syrian reinforcement made its appearance, and in three fierce and successful charges, contributed greatly to damp the ardour of the Persians. From this well-timed assistance, the Day of Succour obtained its name. It was signalized by the heroism of a Mohammedan warrior, Mahujen, who, for having indulged too freely in the use of wine, was doing penance by order of his general in one of the chambers of the castle. Seeing the battle raging below, he could not restrain his enthusiasm; and having prevailed with one of the female attendants, Selma, to undo his fetters, under promise of returning in the evening, he mounted a piebald charger, and sallied into the field. His extraordinary prowess was instrumental in securing the victory; and as his person had not been discovered, he resumed his chains and his captivity, in strict conformity with his engagement. Saad, who had witnessed the seasonable gallantry of the intrepid stranger, was surprised to find himself indebted to the interposition of his own prisoner; he embraced him with great affection; presented him with the mare and the armour he had used with so much distinction; released him forthwith from his confinement, and from all restriction in future with regard to his favourite indulgence. The loss of the Moslems in this single action is stated at 2000, and that of the Persians at 10,000 men.

On the third morning, the contest was again commenced; and if we may conjecture from the title of the Day of Cormorants, the carnage must have exceeded that of the preceding. The Arabs shouted one universal tecbir to terrify the enemy; yet such was the desperate pertinacity with which both sides maintained their ground, that "when the ministers of destiny," to use the flowery language of the East ern historians, "had conducted the chariot of the sun to the obscure chambers of the West, the battle still raged, with unabated fierceness, by the light of their flambeaux, during the whole of the succeeding night; the cupbearers of death busily paraded the bloody field with remorseless rapacity, administering to the unfortunate the bitter draught of dissolution, while the stars in the enamelled vault of heaven continued to witness the sanguinary tumult till the harbingers of the morning announced the return of day." This nocturnal conflict received the whimsical though descriptive name of the Night of Barking; from the discordant clamours of the troops resembling the inarticulate sounds of ferocious animals.

Neither fatigue nor want of rest could slacken their exertions, till near the noon of the fourth day, when a real or imaginary whirlwind drove a cloud of dust against the faces of the infidels. It bore away the pavilion or canopy under which Roostum, on his bed of state, was viewing the progress of the action; and so impetuous were the heat and the tempest, that he was compelled to take shelter among the baggage-mules. The ranks of the Persians were soon thrown into disorder, and attacked by the Arabs, better accustomed to the hurricanes of the Desert. The empty throne arrested their attention; abandoned by its master, who was detected behind one of the beasts of burden. The beauty of his tiara, and the surpassing richness of his girdle and mail, proclaimed the prize which fortune had cast in their way. The danger was imminent, and in the hope of escape he threw himself into the rivulet. Hullal instantly dismounted; plunged without hesitation after him into the stream; and, seizing him in the struggle by the heel, he succeeded in making him his prisoner. The victor then ascended the throne, and with the head of the Persian general fixed on his lance, announced to the armies that "the Lord of the Kaaba was triumphant." Galen experienced a similar destiny, being overthrown and slain in the flight. The Moslems confess a loss of 7500 men, and reckon that of the enemy at 100,000. There may be partial exaggeration in the narrative and numbers of this famous battle; but one thing is certain, it determined the fate of Persia.

Saad appropriated the spoil, the magnitude and value of which excited the admiration of the conquerors. To Hullal he assigned the costly habiliments of Roostum; the tiara alone was estimated at 100,000 dinars (46,250*l.*), and his girdle at 70,000 (32,375*l.*). The armour of Galen was adjudged to Zohara, who received an additional sum of 30,000 dinars (13,875*l.*), in exchange for the imperial standard of Persia, which he had the fortune to take. This celebrated banner was originally the apron of Kawah, a blacksmith of Ispahan, whose intrepidity freed his country from the bloody tyrant Zohauk, and raised Feridoon to the throne. It was rich in ornaments, to which every succeeding king had made contributions; and at the time of the Mohammedan conquests, it had increased from its original shape and size to the length of two and twenty feet, by fifteen in breadth, covered with jewels of very great value. These, however, comprised but a small portion of the sumptuous booty of Cadesia, which included among other articles two shields, each reckoned worth 1,000,100 drachms (22,919*l.*).

The ludicrous mistakes of the Arabs show their ignorance of their own good fortune. Camphor, to the name and properties of which they seem to have been entire strangers, they mingled with their bread, mistaking that odoriferous gum for salt, and were surprised at its bitter taste. "I will give any quantity of this yellow metal for a little of the white," said the soldiers of the Desert, who willingly offered gold, which they had never seen, in exchange for silver, the use of which was better known The legal fifth was conveyed to Medina, consisting of treasure beyond computation, jewels inestimable, furniture of gold and silver, brocades and cloths of silk, embroidered caparisons of horses, camels, mules, and arms of every description. The victory of Cadesia was followed by other more rapid and extensive conquests. Saad, in the month of November, crossed the Euphrates, and in a single battle reduced the whole Mesopotamian peninsula. With a force augmented to 60,000 horse, he next crossed the deep waters of the Tigris: the terrified Persians had fled without offering the least opposition, and could not forbear exclaiming that an army of demons was coming upon them. These accumulated losses, and a superstitious belief that the last day of their religion and empire was at hand, induced Yezdijird to abandon his capital of Madayn. Having lodged a considerable part of his treasures in boats on the Tigris, he fled to Jelwallah, at the foot of the Median hills, taking with him his family and the more valuable of his effects. The spiritless troops followed his example, leaving their country at the mercy of the Saracens, who marched onward, shouting in religious transport, as they entered the gates of the deserted metropolis, "This is the white palace of Khoosroo! This is the promise of the apostle of God!"

The invaders could not express their mingled sensations of surprise and delight, while surveying in this splendid capital the miracles of architecture and art, the gilded palaces, the strong and stately porticoes, the abundance of victuals in the most exquisite variety and profusion, which feasted their senses, and courted their observation on every side. Every street added to their astonishment, every chamber revealed a new treasure; and the greedy spoilers were suddenly enriched beyond the measure of their hopes or their knowledge. To a people emerging from barbarism, the various wonders which rose before them in all directions, like the effect of magic, must have been a striking spectacle. We may therefore believe them when they affirm, what is not improbable, that the different articles of merchandise,—the rich and beautiful pieces of manufacture which fell a prey on this occasion,—were in such incalculable abundance, that the thirtieth part of their estimate was more than the imagination could embrace. The gold and silver, the various wardrobes and precious furniture, surpassed, says Abulfeda, the calculation of fancy or numbers; and the historian Elmacin ventured to compute these untold and almost infinite stores at the value of 3,000,000,000 pieces of gold.*

One article in this prodigious booty, before which all others seemed to recede in comparison, was the superb and celebrated carpet of silk and gold cloth, sixty cubits in length and as many in breadth, which decorated one of the apartments of the palace. It was wrought into a paradise or garden, with jewels of the most curious and costly species;

* Gibbon (in a note, chap. li.) ventures to arraign the accuracy of Elmacin, or rather of the Latin version of Erpenius. But the accuracy of the Saracen historian, and his learned translator, is confirmed by Ockley (vol. i. p. 230), the original Arabic being correctly rendered 3,000,000,000 pieces of gold. The pompous arrogance of Gibbon, who confessed himself "totally ignorant" of oriental languages, is rather amusing, in charging with error a man who is celebrated as the restorer of Arabic literature in Europe. "Erpenius felicissimus ille Arabicarum literarum *instaurator*" is the compliment paid him by Hottinger. If we take each of these pieces at the value of a dinar, which in all probability was the price meant, then the whole will be equivalent to 1,387,500,000*l.* sterling, exceeding by 139,159,375*l.* sterling the total value of gold and silver extracted from the mines of America between the years 1499 and 1803, a period of 304 years. But when we take into account the difference in the value of money then and now, the whole produce of all the gold and silver mines on the globe would not amount to that sum in 1000 years.

the ruby, the emerald, the sapphire, the beryl, topaz, and pearl being arranged with such consummate skill as to represent, in beautiful mosaic, trees, fruits and flowers, rivulets and fountains; roses and shrubs of every description seemed to combine their fragrance and their foliage to charm the senses of the beholders. This piece of exquisite luxury and illusion, to which the Persians gave the name of *Baharistan*, or the mansion of perpetual spring, was an invention employed by their monarchs as an artificial substitute for that loveliest of seasons. During the gloom of winter they were accustomed to regale the nobles of their court on this magnificent embroidery, where art had supplied the absence of nature, and wherein the guests might trace a brilliant imitation of her faded beauties in the variegated colours of the jewelled and pictured floor. In the hope that the eyes of the caliph might be delighted with this superb display of wealth and workmanship, Saad persuaded the soldiers to relinquish their claims. It was therefore added to the fifth of the spoil, which was conveyed to Medina on the backs of camels. But Omar, with that rigid impartiality from which he never deviated, ordered the gaudy trophy to be cut up into small pieces, and distributed among the chief members of the Mohammedan commonwealth. Such was the intrinsic value of the materials, that the share of Ali alone, not larger than the palm of a man's hand, was afterward sold for 20,000 drachms (458*l.* 6*s.* 8*d.*), or, according to others, for as many dinars (9250*l.*). Out of this vast store the caliph granted pensions to every member of his court in regular gradation, from the individuals of the Prophet's family to the lowest of his companions, varying from 275*l.* to 4*l.* 11*s.* per annum.

The military part of the booty was divided into 60,000 shares, and every horseman had 12,000 dinars (5550*l.*); hence, if the army consisted of 60,000 cavalry, their united shares would amount to the incredible sum of 333,000,000*l.* sterling. The crown and wardrobe of Khoosroo, richly adorned with jewels, had been removed; but the mule that carried them away was overtaken by the pursuers, and the spoils of the great king were lodged in the treasury of Medina. On this occasion was fulfilled, as the Moslems affirm, a prophecy of their apostle, that Soraka (the hairy veteran who had nearly taken him prisoner on his flight from Mecca) should wear the belt and bracelets of Khoosroo Purveez. The sack of Madayn (or Ctesiphon), which happened the same year with the reduction of Jerusalem, was followed by its desertion and gradual decay. A single arch, supposed to be the entrance to the palace, is the only vestige that remains to indicate the spot where once stood the proud capital of the Sassanides.

The air, perhaps the luxuries, of Madayn, was found to disagree with the constitution of the Arabs. Saad, with some difficulty, obtained leave of the caliph to withdraw to the western side of the Euphrates, where, on the border of their native deserts, he lodged his followers in a cantonment of mats and reeds. Cufa is the name, in the language of Arabia, for a residence constructed of such materials; but after its destruction by fire it rose from its ashes, and became celebrated under its original appellation as the temporary seat of the caliphate. Another city (Bussora), equally famous, was founded by the Arabs at the same time; with a view of intercepting the communication between Persia and the shores of Hindostan.

The Saracens pushed on their conquests with unabated vigour. No fewer than seventy-seven towns were reduced under their yoke, including Susa, the capital of Susiana or Chusistan. But the turbulent spirit of the new colony at Cufa occasioned a change in the generalship of the army. Saad was replaced by Amar ibn Yasser, a name of considerable celebrity in the annals of Islam. His removal imboldened Yezdijird, who had been driven from Jelwallah with the loss of 100,000 men and immense treasure, to make another effort for the recovery of his dominions. Troops drawn from Khorasan, Rhé Hamadan, and those provinces which the spoliation of the enemy had not yet reached, were assembled, to the number of 150,000; the command of whom was assigned to Firoozan, one of his ablest generals. Amar applied to Medina for reinforcements; and Nooman ibn Makran was the person selected by Omar as leader of the Saracen host, which amounted only to 30,000 men.

Nahavund, an obscure town among the hills, 45 miles south of Hamadan (the ancient Ecbatana), was the memorable post where the Persians ventured to make a final stand for their religion and their country. Their position was strong, fortified with a rampart and a ditch; and two months were wasted in a series of partial and ineffectual hostilities. Firoozan made tenders of accommodation, which, as usual, ended in disappointment. On admission into his pavilion, the rude Arab who acted as ambassador, perceiving him seated on a golden throne with a radiant tiara on his brow, and a crowd of officers standing around, insolently declared that until they were masters of the royal emblems that glittered so brilliantly before him, his countrymen would never consent to recede; and without further ceremony, bounding forward, he seated himself on the throne by the side of the Persian satrap. The negotiation ended with the alternative of tribute or battle. The contest was long and obstinate before success declared for either party. In marshalling his troops, Nooman thus addressed them:—"My friends prepare yourselves to conquer, or to drink of the sweet sherbet of martyrdom. I shall call the tecbir three times; at the first, gird your loins; at the second, mount your steeds; at the third, point your lances, and rush to victory or paradise." On the third day, early in the action, this leader fell mortally wounded by an arrow. He had delayed the attack till the afternoon, the favourite hour of battle with the Prophet, at the moment when the supplications of the faithful from every pulpit and mosque were ascending to heaven in aid of their armies; and when he received his wound was in the act of giving the last tecbir. His death was concealed; but a day of terrible slaughter irrevocably sealed the destiny of the ill-fated empire. Thirty thousand Persians were left on the field; 80,000 perished amid the confusion in their own intrenchments; 4000 who fled with Feroozan to the neighbouring mountains were overtaken, and all put to the sword. The loss of the Moslems was great, but the result was decisive; and the triumph at Nahavund, achieved in the twenty-first year of the Hejira is remembered by the Arabs as the *Victory of Victories.* The booty was prodigious, though small when compared with the wealth of the metropolis. In the equipage of the flying general, it is said, was a crowd of mules and camels laden with honey,—an incident that may serve to indicate the luxurious impediments of an oriental army. The unhappy Yezdijird was thunderstruck by this new disaster. He fled from city to city,—his own governors shutting their gates against him,—until at length he fell near Meru, by the hand of an assassin. He was the last sovereign of a dynasty that had governed Persia during 415 years. His daughters were carried into captivity, and given in marriage to the victors. Hassan, the son of Ali, espoused the one, and Mohammed, the son of Abu Beker, the other; and thus the race of the caliphs and imams was ennobled by the blood of their royal mothers.

After the defeat at Nahavund, the Saracen leaders soon overran the whole country as far as the Oxus, destroying with bigoted fury all that was useful, grand, or sacred. Hamadan surrendered on capitulation; and Ispahan, after a brave resistance, was compelled to submit to the prowess of Abdallah. The cities of Shirwan, Rhé, Tabreez or Tauris, Casbin, and Kom were taken, while the provinces of Azerbijan and Mazunderan, comprehending the ancient Media, Armenia, and Hyrcania, fell before the march of the conquerors, who thus extended their victories from the shores of the Caspian to the Mediterranean. Their progress eastward was equally rapid and extensive. By the direction of Omar, 20,000 men invaded the province of Khorasan, which, with those of Kerman, Mekran, Seistan, and the distant Balkh, were added to the possessions of Islam. On the walls of Ahwaz, Istakhar (the renowned Persepolis, and capital of Fars), Kej, Herat, Meru, and other places of importance, the standard of Mohammed was planted by his impetuous disciples. The greater portion of the vanquished, preferring the abandonment of their religion to oppression or death, adopted the faith of their new masters, while the recusants fled self-banished into distant lands. The administration of Persia was regulated by an actual survey of the people, the cattle, and the fruits of the earth,—a monument which attests the vigilance and justice of Omar. With amazing celerity the Arabs had traversed from the Indian Ocean to the Oxus; but these vast acquisitions were not yet perfectly secured. The succeeding caliphs had many formidable insurrections to quell in their Persian dominions. On various occasions, the inhabitants evinced their abhorrence of a foreign yoke, and their regard for the fire-worship of their ancestors; until at length defeat, massacre, and exile quashed the spirit of revolt, and succeeded, with few exceptions, in blending the vanquished with their oppressors under the united and powerful sway of the Koran.*

* "The Arabs in Khorasan, in Balkh, and even in the vicinity of Bokhara are still numerous; but, except in the former, they

CHAPTER IX.

WARS OF THE CALIPHS.

Invasion of Egypt—Reduction of Farmah or Pelusium—Siege and Capitulation of Memphis—Surrender of Alexandria—Burning of the Library—Conquest and Description of Egypt—Assassination of Omar—Accession of Othman—Capture of Cyprus, Ancyra, and Rhodes—Invasion of Africa—Defeat of the Prefect Gregory—Murder of Othman—Accession of Ali—Political Disturbances—Battle of Seffein—Moawiyah, Founder of the Ommiadan Dynasty, usurps the Caliphate—Ali assassinated—Abdication of Hassan—Death of Moawiyah—Fate of Imam Hossein at Kerbela—The Ommiades or Caliphs of Damascus—Their Character—Expelled by the Family of Abbas, who seize the Throne.

EGYPT was a country familiar to the readers of the Koran; but it was a land of romance,—for in their eyes a cloud of superstitious wonders rested on its actual condition. The pride and the fall of Pharaoh, and the prodigies that attended the flight of 600,000 Israelites, were recorded in their sacred volume as warnings to a disobedient people. Of its population, its strong and numerous cities, its stupendous buildings and mysterious river, they had but obscure conceptions, which tended rather to damp than stimulate their adventurous zeal. But the magnanimous Omar had seen the diadems of Cæsar and Khoosroo laid at his feet, and he still trusted to God and the sword. The conquest of this ancient and celebrated kingdom, then in the hands of the Romans (or Greeks

have no chiefs of any distinction, it having been the policy of both the Tartars and Afghans to scatter and weaken them. Though many of these tribes have preserved the name and appearance of Arabians, they have completely lost the language." *Malcolm's History of Persia*, vol. i. p. 146.

as they are indifferently called), was intrusted by the caliph's commands to Amru, whose valour had already been tried in most of the battles and sieges in Syria.

Farmah or Pelusium, the key of Egypt, was reduced after a siege of thirty days; and by this conquest the country was laid open as far as the modern Cairo. Following the course of the Nile, the Saracens marched directly to Misr or Misra, the Memphis of the old geographers, situated on the western bank of the river, fifteen miles above the Delta, and occupying the spot where the village of Ghizeh at present stands. This important city Amru invested with 4000 men; and after a fruitless siege of seven months, he was obliged to solicit a reinforcement from the caliph, who instantly despatched a force of equal amount, under the command of Zobeir. Even this supply, with all their experience in the Syrian wars, would have proved insufficient to carry the place, but for the religious hatred between the Jacobites and the Melchites, and the treachery of Makawkas the governor. This person, though the prefect or lieutenant of Heraclius, was a mortal enemy to the Greeks; and having amassed an immense fortune by embezzling the tribute of Egypt, he resolved to betray the interests of his master in order to secure the vast treasures he had so unjustly accumulated. The city capitulated on the usual terms, each male of age being rated at two dinars yearly. The number of taxable persons Elmacin computes at 6,000,000; consequently the sum annually collected must have amounted to 12,000,000 dinars of gold (5,550,000*l.*).

The Greeks had assembled a considerable force in the Delta, where the natural and artificial channels of the Nile afforded a succession of formidable defences. To that quarter Amru immediately directed his attention; trusting his safety to the zeal and gratitude of the Egyptians, who had promised, in their oath of fealty, a hospitable entertainment of three days to every Mussulman who should travel through their country. His march from Memphis was a series of skirmishes and victories, some of which were obstinately contested; and after two-and-twenty days of general or partial combat the Arabs pitched their tents before the gates of Alexandria. This magnificent city, perhaps the most arduous and important of the Saracen conquests, had risen, after the lapse of 1000 years, to be the second capital of the empire, and the first emporium of trade in the world. When invested by the Arabs it was abundantly replenished with the means of subsistence and defence.

The besieged made a noble and determined resistance; and, for a while, the valour of the Moslems, whetted as it was by the richness of the prize, quailed before the bravery of the garrison. In every sally and attack the sword and the banner of Amru glittered in the van. Perceiving that the troops were greatly annoyed by the incessant discharge of missiles from one of the castles or towers, he resolved to make himself master of the post. But his imprudent courage betrayed him; for, after a warm dispute, the Arabs were repulsed with some loss; their general with his friend and his slave were taken prisoners, and carried before the prefect or governor of the city. Fortunately his person was unknown; but, in remembering his dignity, he forgot his situation. His resolute demeanour, and the air of authority in which he offered the conditions of tribute or Islam, revealed to one of the spectators the lieutenant of the caliph. "Take off his head," said the courtier; and already the battle-axe of a soldier was raised to strike the audacious captive to the ground. The dexterity of his slave Werdan, who understood Greek, saved his master's life. Affecting to treat him as a menial, he commanded him, with an angry tone, and a box on the ear, to be silent and let his betters speak. A ready tale was feigned of a design on the part of the Saracen general to raise the siege and negotiate an accommodation. The credulous Christians were deceived, and the prisoners dismissed; but they had soon cause to repent of their folly; instead of a pacific embassy, their ears were assailed with the tumultuous acclamations of joy in the camp of the Arabs, as they hailed the return of the brave Amru. Wearied and harassed with a siege of fourteen months, they consented, on the first Friday of the new moon of Moharran (Dec. 22, A. D. 640), to the surrender of Alexandria.

The intelligence of this success, which cost the Moslems 23,000 lives, was conveyed to the caliph, with an intimation that the impatient troops were desirous of exercising their right of conquest in plundering the town. The bulletin of Amru may give some idea of its extraordinary wealth and magnitude. "I have taken the great city of the West It is impossible for me to enumerate the variety of its riches and beauty; and I shall content myself with observing that it contains 4000 palaces, 4000 baths, 400 theatres or places of amusement, 12,000 shops for the sale of vegetable food, and 40,000 tributary Jews. The town has been subdued by force of arms, without treaty or capitulation, and the Moslems are eager to seize the fruits of their victory." Omar rejected the idea of pillage, commanded his lieutenant to restrain the impetuous rapacity of the soldiers, and preserve the wealth and revenues of the place for the public service. The inhabitants were numbered; and, besides the rate of two dinars a head, a tax was laid on vineyards, and properties, according to their annual rent.

One event connected with the capture of this city is too famous to be passed in silence,—the burning of the Alexandrian library; which has done more to familiarize us with the name of Omar than all the other devastations committed in his reign. Presuming on his familiarity with Amru, John the Grammarian had ventured to solicit the gift of the royal manuscripts, which he observed the conquerors had omitted, as of no value, in sealing up the other magazines and repositories of wealth. The general seemed disposed to comply; but as it was beyond his power to alienate any part of the spoil the consent of the caliph was necessary. The answer of the latter is well known,—a sophism that might convince an ignorant fanatic, but could only excite the astonishment and regret of a philosopher. "What is written in the books you mention is either agreeable to the book of God (the Koran), or it is not: If it be, then the Koran is sufficient without them; if otherwise, 'tis fit they should be destroyed." The indiscriminate sentence was executed with blind obedience; and thus perished the literary stores which the pride or learning of the Ptolemys had collected from all parts of the world. They were distributed as fuel to warm the baths of the city; and such was the prodigious multitude of volumes, that six months were scarcely sufficient for their consumption

This instance of oriental barbarism is by some doubted, or totally discredited. "For my own part," says Gibbon, "I am strongly tempted to deny both the fact and the consequences." And his skepticism, in which he is joined by Mons.Ianglés, he supports by the silence of the two annalists, Elmacin and Eutychius, who amply describe the military transactions of the Arabs; both Christians; both natives of Egypt; and both anterior to Abulfarage, the only author, he alleges, who records the catastrophe. Were this statement correct, or did the event rest upon "the solitary report of a stranger who wrote at the end of 600 years," it might stagger our faith in the narrative of the Arminian historian. But the fact does not rest on his sole authority. On the contrary, the circumstance is mentioned by Makrisi and Abdollatif, both of whom have written expressly on the antiquities of Egypt.* Taking the account as recorded (for it is impossible to explode it as a fiction invented by Abulfarage), it will not be easy to estimate the loss which literature sustained. It is true, that in talking of the libraries of antiquity, we must not be misled by magnificent descriptions, or the ample catalogues of their contents. The manuscripts were numerous, but the matter they contained would in modern print be compressed within a small space. The fifteen books, for example, of Ovid's Metamorphoses, which then composed as many volumes, are now reduced to a few dozens of pages. Yet we cannot renounce the belief that, though much has fortunately escaped the ravages of ignorance and the calamities of war, a great deal must have perished in the sack of this famed metropolis. The fall of Alexandria may be said to have achieved the conquest of Egypt. A proportion of taxes, for the benefit of the state, was deducted, according to their annual income, from the clear profits of the wealthier classes, and of those engaged in the pursuits of agriculture and commerce.

At no season could the possession of this hostile

* The Baron de Sacy, in a long note to his translation of Abdollatif (Relation de l'Egypte, p. 240), has collected various testimonies from the works of Arabian writers, preserved in the Royal Library at Paris, which concur in establishing the credibility of Abulfarage's narrative. But these the arrogant Gibbon had never seen. Professor White, in his Egyptiaca (p. 56), enters into an indignant refutation of Gibbon's doubts, and shows that his references to Aulus Gellius, Ammianus, and Orosius are foreign to the purpose for which they are cited, as these writers only notice the accidental conflagration of the Alexandrian library, in the time of Julius Cæsar, when 400,000 volumes are said to have been destroyed. A considerable number was saved in the Temple of Serapis, and at the time of the Saracen invasion the collection had increased to 700,000 volumes.—*Enfield's Hist of Philos.* vol. ii p. 227. *Horne's Introd. to the Study of Bibliography.*

territory have occurred more opportunely for the Arabs. Their own country was visited by a famine, and Omar earnestly solicited a supply of corn for the starving inhabitants. The demand was instantly and profusely answered: a train of camels, bearing on their backs the produce of the gardens and granaries of Egypt, was despatched for their relief, extending in a continuous chain from Memphis to Medina, a distance of 100 leagues. The tediousness of this mode of conveyance suggested to the caliph the plan of opening a maritime communication between the Nile and the Red Sea,—an experiment which Trajan, and the Ptolemys, and the Pharaohs had attempted in vain. The resources of the Arabs were equal to its accomplishment; and a canal of at least eighty miles in length was opened by the soldiers of Amru. This inland navigation, which would have joined the Mediterranean and Indian Ocean, continued in use for some time, if we may believe Eutychius and Elmacin, and was called the River of the Emperor of the Faithful. But when the caliphs removed their seat of government from Medina to Damascus, its utility was sacrificed to an apprehension of the danger which might ensue from its opening to the Grecian fleet a passage to the holy cities of Arabia.

The anxiety of Omar to learn something of this new territory, of which he had but an imperfect and legendary knowledge, was natural; and the lively description of Amru in his answer would augment rather than diminish his romantic conceptions of this singular country. "O, Commander of the Faithful! Egypt is a compound of black earth and green plants, between a pulverized mountain and a red sand. The distance from Syené to the sea is a month's journey for a horseman. Along the valley descends a river, on which the blessing of the Most High reposes, both in the evening and morning, and which rises and falls with the revolutions of the sun and moon. When the annual dispensation of Providence unlocks the springs and fountains that nourish the earth, the Nile rolls his swelling and sounding waters through the land. The fields are overspread by the salutary flood, and the villagers communicate with each other in their painted barks. The retreat of the inundation deposites a fertilizing mud for the reception of the various seeds; the crowds of husbandmen who blacken the fields may be compared to a swarm of industrious ants; and their native indolence is quickened by the lash of the taskmaster, and the promise of the flowers and fruit of a plentiful increase. According to the vicissitudes of the seasons, the face of the country is adorned with a silver wave, a verdant emerald, or the deep yellow of a golden harvest."* The phenomenon of a country alternately a garden and a sea was new to the dwellers in the Desert. Their imagination took a license from these wonders; but a more accurate inquiry has enabled us to rectify many of their fabulous and exaggerated statements. Their 20,000 cities and villages are limited, by an authentic estimate of the twelfth century, to 2700; the 20,000,000 of inhabitants have collapsed to one-fifth of that calculation; and the 300,000,000 pieces of gold or silver that were annually paid to the treasury of the caliph have been found to be an error of Elmacin, and reduced by Renaudot to the more moderate revenue of 4,300,000 pieces of gold, of which 900,000 were consumed by the pay of the soldiers.†

The ambition of Amru was not content with a single conquest; he began to extend his victories westward into the kingdoms of Africa, and in a short time made himself master of all the country which lies between the Nile and the desert of Barca.

* Murtadi, Merveilles de l'Egypte.
† Hist. Patriarch. Alex. p. 334

In the midst of this career, and after the rapid acquisition of a power which, had it never extended farther, might deserve the name of a formidable empire, the dagger of an assassin put an end to the life and reign of Omar. This obscure individual was a Persian slave, named Ferouz. Watching his opportunity, while the caliph was engaged at morning prayers in the mosque, the murderer rushed forward to the pulpit, and with a small dagger or khunjer inflicted six wounds on his person, one of which, below the navel, was pronounced mortal. The assembly were thrown into the greatest consternation, and immediately surrounded the villain who had imbrued his hands in the blood of their sovereign. He made a desperate resistance, having wounded thirteen of his assailants, seven of them mortally. At length he was seized by one of the caliph's attendants, who threw a mantle over him: but he avoided the torture that awaited him, by stabbing himself on the spot.

The piety, justice, abstinence, and simplicity of Omar procured him more reverence than his successors with all their grandeur could command. "His walking-stick," says Alwakidi, "struck more terror into those who were present than another man's sword." His diet was dates, or coarse barley bread dipped in salt; his drink, water; and sometimes, by way of penance, he would eat his bread without salt. Of religious duties he was a punctual observer, and, during his brief caliphate, had performed nine pilgrimages to Mecca. He preached in a tattered cotton gown, torn in twelve places; and a Persian satrap, when he paid homage to the conqueror of his nation, found him lying asleep among the beggars on the steps of the mosque at Medina. During the ten years of his administration, 4000 churches were destroyed, and 1400 mosqes erected on their ruins.

Omar had devolved on six companions the arduous task of choosing a Commander of the Faithful The conclave assembled without delay, and for a while the balance of the Moslem succession trembled between Othman and Ali. An insidious proposal of Abdalrahman decided the contest: That whoever was chosen should consent to rest the basis of his government on the law of the Koran and the traditions, and *the virtuous example of his two predecessors.* This latter clause Ali rejected, as it implied a tacit acknowledgment that he had no inherent or preferable title to the throne. Othman, who was afflicted with no such scruples, embraced the terms without limitation or restriction.

Partial revolts, both in Egypt and Persia, threatened to disturb the commencement of his reign; but they were suppressed by the prompt application of the sabre. The Greeks on the frontier of Asia Minor, and the Tartars in Shirwan, and on the western shore of the Caspian, were compelled to submit to the yoke of Islam. In the Levant the arms of the Arabs were equally successful. Moawiyah, the lieutenant of Syria, equipped a powerful armament, and invaded the island of Cyprus. As neither the inhabitants nor the imperial troops were able to oppose him, they agreed to capitulate, on condition that the whole revenues of the island, amounting to 14,400 ducats, should be equally divided between the caliph and the Emperor Constance, grandson of Heraclius. This tribute the Saracens enjoyed for nearly two years, when they were dispossessed by the Christians. On the reduction of Cyprus, Moawiyah took Ancyra and Aradus (Ruad); a rocky islet, 200 paces in the sea, but containing a populous city and a fortress of great strength. Penetrating as far as Rhodes, he seized that island, and destroyed the famous brazen Colossus, a gigantic statue of Apollo or the Sun, seventy cubits (105 feet) in height, erected across the entrance of the harbour, to commemorate the fruitless siege of Demetrius, about three centuries before the Christian era. The massy trunk and huge fragments were carefully collected by the mercenary Saracens, and sold to a Jewish merchant of Edessa, who loaded 900 camels with the metal, the value of which has been estimated at 36,000*l.* sterling money. Exclusive of the immense booty acquired on this expedition, the Arabs are said to have carried off 8000 beautiful captives of both sexes.

In Egypt the affairs of the caliph suffered a temporary reverse, occasioned entirely, however, by his own imprudence in displacing the popular governor. But the restoration of Amru re-established the Saracen ascendency. The Grecian troops were defeated after a protracted combat of several days. Alexandria was again taken by storm, its towers dismantled, and its walls thrown down. By the seasonable interposition of their general, however, the fury of the relentless Moslems was stayed; and the Mosque of Mercy, erected on the spot, commemorated for ages the clemency that arrested the indiscriminate slaughter of the citizens. The attention of Othman was next turned to the reduction of Africa.

At the head of a formidable expedition, amounting to 40,000 men, Abdallah, governor of Egypt, passed the desert of Barca. A painful march brought them under the walls of Tripoli, the well-known capital of a province which still maintains a respectable rank among the States of Barbary. The fortifications of this place resisted the first assaults of the invaders. A detachment of Greeks was surprised and cut to pieces on the seashore; but the approach of Jujeir (Gregory), the prefect or lieutenant of the emperor, with a force of 120,000 men, consisting of Roman troops and Moorish auxiliaries, induced the Arabs to suspend the siege, in the hope of deciding the fate of the country by a pitched battle. Gregory rejected with scorn the terms of the Saracen general. In the midst of a sandy plain the two armies were fiercely engaged for several days, from morning light till the hour of noon, when heat and fatigue compelled the soldiers to seek the shelter of their tents.

By the side of Gregory fought his daughter,--a maid of surpassing beauty and spirit. The richness of her armour and apparel was conspicuous in the foremost ranks of the battle. Her charms had captivated every heart, and the courage of the bravest was excited to a desperate enterprise by the promise of her hand and 100,000 pieces of gold to the man who should bring the head of Abdallah to the camp of the Africans. The prize was glorious; but, unfortunately, it offered an equal temptation to the enemy, which the genius of Zobeir, who had signalized himself in Egypt, converted into a successful stratagem. He advised the Moslem general to retort on the infidels their ungenerous attempt, and proclaim through the ranks that the head of Gregory should be repaid with his captive daughter, and her magnificent dower.

To the courage and discretion of Zobeir was intrusted the execution of his own stratagem. Both armies engaged next morning as usual, and the combat was prolonged till the heat of the mid-day sun became insupportable. The Greeks retired to their camp; the Mussulmans threw down their swords, unbridled their horses, laid their bows across their saddles, and by every appearance of lassitude deceived the enemy into security. On a sudden the charge was sounded; Zobeir, with a part of the forces who had lain concealed in their tents, sprang from their ambush, fresh, active, and vigorous; and, mounting their horses, assailed the Greeks, astonished and exhausted with fatigue. The latter seized their arms in haste; but their ranks were soon broken and dispersed. Gregory was slain by the hand of Zobeir; and the scattered remains of his army sought refuge in the town of Sufetala (150 miles to the south of Carthage), which surrendered after a short resistance, yielding a spoil of 3000 dinars of gold (1387*l.* 10*s.*) to every horseman, and 1000 to every foot-soldier.

The prefect's daughter had animated with her courage and her exhortations the soldiers of her country, and sought revenge for the death of her father, till she was surrounded by a squadron of horse, and led captive to the presence of Abdallah. The victor appeared not to claim his precious reward; and from his silence it might have been presumed that he had fallen in the action, until the tears and exclamations of the prisoner at the sight of Zobeir revealed at once the man who slew her father, and the modesty of the gallant soldier. "Why do you not claim the meed of your valour?" said Abdallah, astonished at his indifference in the presence of so much beauty. "My sword," said the enthusiast, coolly, "is consecrated to the service of religion. I fight for a recompense far above the enjoyments or the riches of this transitory life." The virgin and the gold he accepted with reluctance; but a reward more congenial to his temper was the honourable commission of announcing to the caliph the success of his arms. In the mosque at Medina, the companions, the chiefs, and the people assembled, to hear from Zobeir the interesting narrative of the African war; and as the orator forgot nothing except his own merits, the name of Abdallah was joined by the Arabs with those of Khaled and Amru. Discouraged by these reverses, the natives appear to have relinquished all hope of further resistance, and their conqueror consented to grant them a precarious peace for the almost incredible sum of 2,500,000 dinars of gold (about 1,166,666*l.* 13*s.* 4*d.*).

The death of Othman, and the political feuds that distracted the reign of his successor, suspended the progress of the western conquests of the Arabs for nearly twenty years. A spirit of discontent had begun to prevail generally throughout his dominions, and this was aggravated by a continued system of favouritism on the one hand, and of ill-judged severity on the other. The malecontents in the different provinces held correspondence on their mutual grievances and the means of redress. To appease their fury, Othman owned from the pulpit of the mosque the faults of his administration.

But he soon forgot his resolutions; and when crowds of the factious nobles and indignant citizens assembled at the gates of the palace, he refused to grant them either audience or admission. A fierce and sanguinary contest ensued; for a band of 500 of his attendants had rallied in the principal court. At length the assailants made their way to the caliph's apartment, where they found him with a copy of the Koran in his lap. Nine wounds were inflicted on his body, of which he expired. The faithful Nailah threw herself between her husband and the daggers of the assassins, and had the fingers of one hand struck off in this heroic effort of conjugal tenderness.

The charities of Othman were carried to extravagance, yet he left immense riches behind him,—500,000,000 drachms and 150,000 dinars being found in his coffers; besides 200,000 dinars set apart for pious uses (in all, 11,620,208*l.* 6*s.* 8*d.*).

Considering the claims and personal merits of Ali, his affinity to the Prophet, and his tried valour in many a field under the banners of the Koran, it is apparently wonderful that the nation of the faithful should have quietly tolerated his repeated disappointments. Though still opposed by the faction of Ayesha, his implacable enemy, the ostensible obstacles to his elevation were now removed, and the sovereignty had descended to him even in that order of succession which had hitherto been determined rather by degrees of sanctity than of kindred. In assuming the regal and sacerdotal duties he changed not the accustomed simplicity of the caliphs. On the day of his inauguration he went to the mosque dressed in a thin cotton gown tied round him with a girdle, a coarse turban on his head, his slippers in one hand, and his bow in the other instead of a walking-staff. The chiefs and the people saluted their new monarch, and gave him the right hand of fealty and obedience.

But his accession was only the beginning of political convulsions, and the cause of that religious schism which rent the creed of Islam in twain. The discontented faction took the name of Motazalites, or Separatists; and the sinews of war were supplied from the plundered treasury of Yemen. The spirit of discord was irritated by the calumny, diligently propagated and believed, that Ali was an accomplice in the murder of Othman. Resolved to usurp the government of Irak, Telha and Zobeir, the favourite partisans of Ayesha, and 3000 of the insurgents, departed from Mecca for Bussora, with the mother of the faithful at their head. Crossing the Desert of Arabia, they reached the banks of the Euphrates, where a large reinforcement from Syria increased the army of the rebels to 30,000 men.

Intelligence of this daring insurrection was conveyed to Medina. Ali with difficulty mustered a force of 20,000 Arabs; and having advanced to Bussora, he gave battle to the rebels, after vainly attempting a peaceful adjustment of their differences. The struggle was fierce and sanguinary; but victory declared for the caliph. Telha and Zobeir were slain, and Ayesha herself taken prisoner. Her equipage was a kind of litter or cage, secured on every side by strong panels, and fixed on the back of a camel of extraordinary size and speed, and covered with mail. Her shrill voice animated the troops, and her post was in the midst of danger. Her litter was pierced with innumerable darts and javelins, resembling the quills of a porcupine. The bravest in the ranks successively held the reins of the animal, chanting pieces of poetry; and in this arduous duty not fewer than 230 were numbered among the killed and wounded, most of whom had lost a hand. The venerable heroine was not included in the list of bondwomen, but dismissed with a select escort to her home at Medina. The battle of Khoraiba, or the Day of the Camel, the first that stained the arms of the Moslems with civil blood, was fought in the thirty-sixth year of the Hejira (December, A.D. 656).

Master of Irak, Ali received the submission of Egypt, Arabia, Persia, and Khorasan; but the sword of rebellion was not broken. A more formidable adversary remained to be vanquished in Syria, in the person of the governor, Moawiyah, who openly threw off his allegiance, and had been proclaimed caliph by the western provinces. The mosques resounded with cries of vengeance for the death of their legitimate sovereign. To inspire a just abhorrence of the guilty perpetrators, the bloody shirt of the murdered Othman, with his wife's finger pinned upon it, was exposed on the pulpit of Damascus, and paraded at the head of the troops.

The pacific overtures of Ali were rejected; both parties appealed to the sabre as the arbiter of their contested titles; and on the plain of Seffein, which extends along the western bank of the Euphrates, near Racca, the two armies pitched their camps.

On this spacious and bloody theatre 150,000 Moslems waged a desultory war of 102 days; for both competitors seemed unwilling to peril their cause on the hazard of a general engagement. Ninety actions or skirmishes are recorded to have taken place; and in these the humanity of Ali was as conspicuous as his valour. He strictly enjoined his troops invariably to await the first onset of the enemy—to spare the fugitives, and respect the virtue of the female captives. Not a day passed in which he displayed not some extraordinary feat of personal strength and skill. The bravest leaders of the Syrian host fell in succession by the single prowess of his resistless arm,—"For death itself dwelt on the point of his spear, and perdition in the hilt of his sword." The hideous and gigantic Kerreib, who could obliterate with his thumb the impression on a silver coin, he cleft at one stroke from the crest to the saddlebow. Two warriors attacked him in disguise; but with a sweep of his double-scimitar he bisected the foremost through the middle with such rapidity and precision, that the rider remained fixed on the saddle; the spectators concluding he had missed his blow, until the motion of the horse threw the body in halves to the ground.

The two concluding days were the most sanguinary; and such was the carnage, that the most hardened veterans were seen to weep aloud on beholding the scene of destruction around them. On the first charge the ranks of the Syrians were broken. The battle raged nevertheless, the cavalry dismounting to combat on foot, and even on their knees, with their swords and daggers. The contest suffered no interruption from the setting sun. Mounted on a piebald horse, and clad in the accoutrements of the Prophet, with his ponderous sabre, Ali cheered on his troops amid the confusion and obscurity of the night. As often as he smote a rebel, the shout of Allah akbar rose; and before morning he was heard to repeat that tremendous exclamation 523 times.

The usurper of Damascus already meditated flight; for 7000 of his troops had perished in the slaughter of that memorable night: but a stratagem of Amru snatched the victory from the grasp of his rival. He ordered the Syrian soldiers to fix their copies of the Koran, to the number of 550, on the points of their lances, and, stationed in front of the line, to await the approach of the assailants. The artifice was successful; the conscience of the Moslems was overawed by this solemn appeal to the sacred volume. Ali was not to be imposed on by this specious device—but he was compelled to yield to the clamours of his followers, weary of bloodshed, and willing to embrace any reasonable terms of accommodation. "The son of Henda has vanquished!" he exclaimed in an agony of vexation, when he found discord and disloyalty spreading among the fanatical part of his army, who demanded that the sword, already waving in triumph, should return to the scabbard in reverence for the Koran. The battle was accordingly suspended, and Moawiyah saluted caliph. With sorrow and indignation Ali retreated to Cufa; but his partisans separated into factions, and his interest from this fatal period began greatly to decline.

Moawiyah took revenge for the losses he had sustained at Seffein. His troops plundered several districts in Mesopotamia, and for a time maintained possession of Bussora. Other detachments made incursions into Arabia, where they committed terrible devastations. Penetrating into Hejaz, they reduced Medina, Taïf, and Mecca, and extended their ravages into Yemama and Yemen, the garrisons of which, unable to oppose the invaders, fled with precipitation to Cufa. The authority of the usurper was thus quietly established over the sanctuaries of the Moslem faith, and many of the most important provinces of the Saracen dominions.

The same tragic fate awaited Ali, now harassed and heart-broken with misfortune, that had cut off his two immediate predecessors. Three of the Korajites or schismatics, heated with fanaticism and revenge, resolved in secret to expiate the slaughter of their comrades, by plunging their daggers in the bosoms of the three principal leaders, Ali, Moawiyah, and Amru. Each of the assassins poisoned his weapon, and selected his victim. Moawiyah was stabbed in the groin, but the wound was not mortal. Amru was preserved by a fit of indisposition which prevented him from officiating as imam in the mosque. A secretary received the fatal stroke that was intended for his master. The third of the conspirators had better success; and, in the mosque at Cufa, Ali received a blow from his hand, of which he expired in the course of four days.

This prince united the qualifications of a poet, an orator, and a soldier; for he was the bravest and most eloquent man in his dominions. A monument of his wisdom still remains in a collection of precepts or sentences, of which 169 have been translated by Ockley. Many other maxims and poems have been ascribed to him; but some hesitation must be allowed in fixing their authorship. The eulogies of his partisans are fulsome and extravagant: "the king of men—the lion of God—the distributor of lights and graces," are among the epithets which their adoration has conferred on him. During the caliphate of the house of Ommiah the place of his interment was kept concealed. In the fourth age of the Hejira, when the Abbassides ascended the Moslem throne, it was discovered; and a tomb, a temple, and a city arose on the spot—known in modern times by the name of Meshid Ali, five or six miles from the ruins of Cufa, and 120 to the south of Bagdad. The monarchs of Persia have enriched it with a succession of spoils, and thousands of the Sheahs pay their annual visits to the holy sepulchre.

Hassan inherited his father's piety, but he was deficient in courage, and in every qualification necessary to rule a turbulent people. The fickleness and infidelity of his adherents appear to have divested him of all relish for the splendours of royalty; and without deliberation or delay he transmitted a letter to Moawiyah, offering to resign the sovereign power into his hands, on condition that he should enjoy certain revenues; and that no successor to the throne should be appointed during his life. The Syrian chief accepted the terms with the utmost satisfaction. The unambitious prince, weary of the world, retired to Medina, devoting his life to religion, and spending his vast revenues of 150,000*l.* a year in deeds of charity, the whole of which he twice distributed among the poor. Here he fell a victim to the jealousy of Moawiyah; and the criminal deed was perpetrated by his own wife Jaidah (A. D. 670), by rubbing him while warm with a linen cloth impregnated with poison.

The abdication of Hassan left an undivided, but not a powerful, throne to Moawiyah. This prince, the first of the dynasty of the Ommiades, was the son of Abu Sofian, who had usurped the power of Abu Taleb at Mecca, and third in descent from Ommiah, the founder of the family of that name, who was a nephew of Hashem, being the son of a younger brother, and consequently a collateral branch of the noble tribe of the Koreish. His claim, however, according to the principles of legitimacy, was posterior to that of the descendants of Fatima, and even of the children of Abbas, the uncle of the apostle. It is a singular reflection, that the earliest and most inveterate persecutors of Mohammed should have usurped the inheritance of his children, while his person and sanctity were yet fresh in their memories; and that the boldest champions of idolatry should, in the short space of forty years, have become the sovereigns of that hierarchy which they had laboured to overthrow. In his youth, Moawiyah had been honoured with the offices of almoner and secretary to the apostle, and was employed to register his revelations. For twenty years he had exercised supreme authority in Syria, when his elevation to the caliphate divulged the important truth, that the city of the Prophet was not the only place that enjoyed the rights and privileges of election.

His accession did not quiet the murmurs of faction, and his reign was occupied with little else than deposing chiefs and governors, whose loyalty he could not trust; and heaping honours and emoluments on others, whose passions could thus be made subservient to the advancement of his own interests

The intrigues of Ayesha at Medina were no secret at the court of Damascus; and in a personal visit the caliph received an insulting refusal of her allegiance. But his apprehensions from this quarter were speedily removed. The illustrious widow was invited to an entertainment; the chair destined for her reception was placed over the mouth of a deep well or pit, slightly covered with leaves; and the moment the unsuspecting guest seated herself a table, she " sank to everlasting night." Such, say the Persian authors at least, was the fate of this ambitious mother of the faithful. Moawiyah soon after " quitted this abode of clay for the mansions of eternal retribution." The most important feature in the government of this caliph was his changing the monarchy from being elective, and making it hereditary in his own family. Such was his influence, that he succeeded in obtaining, not only the submission of all classes in Syria and Irak, but the acquiescence of Mecca and Medina, to the nomination of his son Yezzid, a feeble and dissolute youth. This prince had the wisdom to retain his father's lieutenants in their places; and his reign was almost exclusively employed in subduing the refractory partisans of Hossein, brother of Hassan.

The melancholy fate of this imam deserves a place in the history of his nation. The injustice done to his fàmily, and the weak character of the reigning caliph, revived a sympathetic loyalty for the child of Fatima, and suggested the thought of reinstating him on the throne. The inhabitants of Cufa invited him to make his appearance in their city; and a list was secretly transmitted to Mecca of 140,000 Moslems in Irak, who professed their attachment to his cause, and were ready to draw their swords as soon as he should arrive on the banks of the Euphrates. Under a slender escort of 40 horse and 100 foot, he left Mecca; and with a numerous retinue of women and children, including the whole of his own family and the greater part of his brother's, he traversed the deserts of Arabia, in the hope of reaching his friends before the lieutenant of Yezzid should have received information of his design. His expectations were miserably disappointed. Obeidallah, the governor of Cufa, had detected and put to death his faithful agent; and, in quenching the first spark of revolt, the defection or ruin of his party was accomplished. As Hossein approached the confines of Irak, the hostile face of the country, the wells and places of refreshment on the roads being destroyed, told the melancholy tidings; and his fears were confirmed by the intelligence that 4000 of the enemy were on their way to intercept him. "Alas," said he, "encumbered with all this family, how can I retreat?" and, quitting the direct route, he pitched his tents by the brook of Kerbela. Here he was immediately surrounded; his attempts to obtain honourable conditions of peace, or a return to Medina, were abortive; for the command of the inexorable Obeidallah was peremptory: "Bring me either Hossein or his head!"

His little band, true to his fortunes and resolved to share his fate, drew up to meet their assailants. Terrified by the disparity of numbers and the certain prospect of death, the women and children gave vent to their sorrows in loud and bitter lamentations. The archers galled them with their arrows, and in one charge twenty of them were killed on the spot. The survivors maintained the combat with unshaken constancy, until the heat of the day and the impulse of despair rendered their thirst insupportable: but relief could not be had; for they were cut off from all communication with the river. The cavalry were entirely dismounted; but they fought on foot, generously throwing themselves between their leader and the swords of the enemy; and each saluting him, as they passed in succession to the deadly encounter,—" Peace be with thee, thou son of the Prophet of God! Fare thee well!" Their only respite was the hour of prayer; and Hossein beheld with tears the last of this band of martyrs expire by his side. The next victims that offered themselves to the unequal conflict were his five brothers; but it was only to perish with their slaughtered companions. His eldest son sought revenge in the thickest of the battle, and fell after bravely sustaining ten different assaults. Hossein was overpowered with feelings of anguish, which he could no longer suppress. Alone, weary, and wounded, he seated himself at the door of his tent, addressing his supplications to Heaven. His infant child was brought to his arms, and while pressing it to his bosom an arrow pierced the unconscious innocent through the heart. His little nephew, while running to embrace him, had his head struck off with a sabre. He was himself wounded in the mouth with a javelin, while quenching his thirst with a drop of water. As the soldiers surrounded the object of their vengeance, his sister Zeinab, in a transport of horror, rushed from the tent, and adjured their general not to suffer the grandson of their Prophet to be murdered before his eyes. Frantic with rage, Hossein threw himself into the midst of them, and the boldest retreated before his desperate charge. Awe kept them at bay until their cowardice was reproached by the remorseless Shamer, a name still detested by the faithful; when the son of Fatima was despatched with three-and-thirty wounds.

The remains of the slaughtered martyrs lay three days unburied on the sands of Kerbela, when they were at length collected by the neighbouring villagers, and committed to one common grave. The wretched female captives were carried to Cufa, thrown across the backs of camels. Thence they were conducted in chains, and entirely naked, to the Syrian capital, and afterward dismissed to Medina. The head of Hossein was laid before the caliph; and, like Obeidallah, he could not refrain from offering the same brutal indignity of beating it on the mouth with a whip or cane, on which an aged Mussulman exclaimed,—"Upon these lips I have seen the lips of the apostle of God." The memory of this imam is still dear to his Persian votaries. The Day of Hossein is an anniversary of weeping and lamentation; and the hatred of the two nations is preserved and prolonged by this solemnity. The first sultan of the Bowides reared a sumptuous monument on the spot (Meshid Hossein), which still marks the sepulchre of the martyr. Crowds of pilgrims continue to pay their annual devotions at his shrine. The soil near his grave is purchased with avidity, and is said to possess the most miraculous virtues; while the privilege of being interred near the remains of the saint is bought by the rich at an extravagant price. Ali, his two sons, and their descendants, to the ninth generation, form the twelve imams of the Sheahs. The last of these, called he *Mohadi* or guide, they suppose to be still alive, but invisible; and that he will appear before the last day, to establish a sort of millennial reign on earth.

The last action of the reign of Yezzid was to quell the insurrections which had broken out at Medina and Mecca, partly on account of Hossein's death, and partly in consequence of changes in the government. The former place was reduced to obedience in a single battle (A.D. 682). The insurgents fled, and took shelter within their walls; but the troops of Damascus entered with them; and for three days the city was one universal scene of pillage and slaughter, during which 6000 of the principal inhabitants are said to have perished. At Mecca, the partisans of Hossein had proclaimed Abdallah, the son of Zobeir, caliph of Arabia, Irak, Khorasan, and Egypt. To this place the Syrian army directed their march; and for forty days it defied all the arts and engines of the besiegers. The walls were severly battered; several pillars of the temple were demolished by stones thrown from the catapults; and one of the machines, loaded with naphtha, being discharged against the Kaaba, the linen veil which covered it was set on fire and reduced to ashes. The city shortly after was stormed by Hejaje: but Abdallah, though deserted by his followers to the amount of 10,000, disdained to surrender, and, rushing furiously into the thickest of the enemy, he obtained from their swords the honourable death which he sought. Before this occurrence had taken place, Yezzid, to use the words of the Persian original, "was enrolled among the dignitaries of the infernal regions." The domestic history of his successors merits little attention. The recovery of Mecca had established the sovereignty of the house of Ommiah over the whole Moslem world; and most of the subsequent reigns were occupied in subduing refractory provinces or quelling the revolts of petty chieftains. Some of these insurrections were attended with extraordinary scenes of atrocity and bloodshed. It was the boast of Moktar, who supported the interests of Abdallah in Persia, that he had destroyed nearly 50,000 of the enemies of the Alides, exclusive of those who perished in the several battles which he fought. Hejaje, who was appointed governor of Hejaz, was a cruel and sanguinary tyrant. During the twenty years that he had presided over Irak, he is said to have destroyed 120,000 persons of rank; and in the various prisons under his jurisdiction were found at his death 30,000 men and 20,000 women

The following Table will exhibit, in the order of succession, the names and reigns of the caliphs who filled the Moslem throne from the death of Mohammed till the termination of the Ommiadan dynasty:

A. D.	A. H.		A. D.	A. H.	
632	11	Abu Beker	684	65	Abdolmalec
634	13	Omar I.	705	86	Walid I.
943	23	Othman	716	97	Soliman
655	35	Ali	718	99	Omar II.
660	40	Hassan	721	102	Yezzid II.
661	41	Moawiyah I.	723	104	Heschain
676	60	Yezzid I.	742	125	Walid II.
683	64	Moawiyah II.	743	126	Yezzid III.
683	64	Abdallah	744	127	Ibrahim
683	64	Merwan I.	744	127	Merwan II.

The hereditary line of the Ommiades was broken in the person of Merwan I., who transferred the caliphate to his own son, instead of leaving it to the younger brother of Moawiyah, II., to whom he had been appointed guardian. The reign of Abdolmalec was signalized by the reduction of the precious metals to a specific standard throughout the empire; and for the establishment of a regular coinage both in gold and silver. Previous to this time the Arabs had adopted no currency of their own, being in the habit of trading with the Greek and Persian money The value of the gold dinar was then reckoned equivalent to 8*s*. sterling. Walid I. abolished the Greek language and characters, which before his elevation had been used in keeping the accounts of the public revenue, and ordered his clerks and secretaries to substitute the Arabic,—a change to which, very probably, we owe the invention, or at least the familiar use, of our present numerical figures; and thus, as Gibbon remarks, a regulation of office has promoted the most important discoveries of arithmetic, algebra, and the mathematical sciences. The leisure time of this prince was occupied in decorating his capital and erecting religious edifices on which he expended vast sums of money.*

Soliman, brother to the preceding caliph, was a mild and liberal prince, but remarkable for his voracity. He died of pleurisy, or indigestion,—an event not at all extraordinary, if what the Arabs assert be true, that he used to devour 100 pounds weight of meat every day, and could dine very heartily in public after eating three roasted lambs for breakfast His favourite dish was the baked intestines of sheep, —thirty of which, with as many cakes of bread, he was frequently known to despatch at a single meal. In poverty and in the economy of his habits, Omar II. imitated and even surpassed his illustrious namesake. The subsistence of himself and his household was limited to the frugal expenditure of two drachms a day. One of his generals, who visited him in his last sickness, found the emperor of the faithful stretched on a couch of palm-leaves, in a dirty undress, supported by a pillow of sheepskins, and covered with a coarse garment. His wife apologized for the sordid condition of his linens, by stating that the imperial wardrobe comprehended no more than a solitary shirt. After his death, his palace was ransacked, in the hope of finding immense treasure; but nothing was discovered except a thick riding-coat, and a rope on which he swang for recreation after his spirits had been exhausted by long and fervent prayer. The reign of Yezzid II. was

* He adorned the temple of Jerusalem with additional buildings, and enlarged that at Medina, by demolishing the low cottages formerly inhabited by the Prophet's wives. But the most magnificent of these structures was the famous mosque of the Ommiades at Damascus, which employed 12,000 workmen, and cost 6,000,000 dinars of gold (2,775,000*l.*), exclusive of the valuable materials contributed by foreign princes. The floor was of white marble, the walls were covered with rich mosaic, and the roof overlaid with gold. The golden chains for the lamps, of which there were 600, were so brilliant as to dazzle the eyes of the worshippers

disturbed by a rebellion in Khorasan; but the progress of the insurgents was arrested by his brother the brave Moslemah, who dispersed them in a battle fought at Hira. He died of grief at his favourite residence on the banks of the Jordan. While amusing himself in the garden by tossing grapes into the mouth of a favourite concubine, one of them unluckily stuck in her throat, and produced immediate suffocation,—an accident which he survived only fifteen days. The life of Walid II. was one continued scene of debauchery, being addicted to gaming, gluttony, drunkenness, and every species of obscenity; and in the pilgrimage to Mecca he carried with him wine and dogs for his amusement.

The names of Yezzid III. and Ibrahim need only be mentioned as among the ephemeral sovereigns who exhibited a precarious and fitful splendour just before the power of the Ommiades was finally extinguished. Merwan II., the fourteenth and last of the race, had signalized himself by his victories in the neighbourhood of the Caucasus, and against the Tartars. His valour and intrepidity gained him the title of *Himar el Jezirah*, or the Mesopotamian Ass,—a name which, however contemptible it may sound in European ears, was a proverbial compliment in Arabia; the warlike breed of that country, instead of an emblem of stupidity, being remarkable for courage, and never known to fly before an enemy. On his accession, he found the whole empire, from a variety of causes, pregnant with insurrection. Emesa set the first example of revolt; the Damascenes next threw off their allegiance; the citizens of Cufa deposed the Syrian governor, and proclaimed a caliph of their own. At Bussora, another rival was elevated to the throne, who supported his pretensions with a numerous army; but he was routed in the neighbourhood of Damascus, with the loss of 30,000 men.

Disgusted with the sanguinary despotism of the

reigning family, the partisans of the house of Abbas had turned their eyes and their wishes towards the illustrious line of Hashem, in preference to the Fatimites, who were passed by as rash or pusillanimous. Mohammed, the grandson of Abbas, in his obscure residence in Syria, had long fostered the secret hopes, and received the homage of his friends. At his death, the rights and honours of his family had been transferred to his son Ibrahim; and when the revolt of Kerman gave them an opportunity of openly disclosing their views, a body of warriors gathered round his person, and the two black standards, called the Night and the Cloud, were unfurled. Under these allegorical emblems, Abu Moslem proclaimed his master the rightful emperor and imam of the faithful. Party colours have always been used as badges of distinction among contending factions. The Moslems adopted this expedient in the approaching war of succession; green was the symbol of the Fatimites; white of the Ommiades; while black was adopted by the followers of Abbas. Cities and provinces hoisted the respective flags; and from the Indus to the Euphrates the Eastern world was convulsed between the struggles of the black and the white armies. Even the service of the mosque was divided between the rival sovereigns; some reciting their prayers in the name of Merwan, others in that of Ibrahim. In Korasm, the interests of the Ommiades had been bravely maintained by the governor, Nasr Sayar; but a series of disasters and defeats at length convinced him that he was no longer capable of contending with the invincible Abu Moslem. Ibrahim was unfortunately waylaid in a pilgrimage to Mecca, seized, and carried to the presence of the caliph, who forthwith ordered him to be suffocated by enclosing his head in a bag of quicklime. His two younger brothers, Saffah and Almansor, happily eluded the search of the tyrant, and escaped to Cufa, where the former was saluted caliph from the pulpit. But it was in the battle-field rather than in the mosque that the important controversy was decided. Already Abu Moslem and his lieutenant, Kotaba, had reduced the western provinces of Irak, from the Euphrates to the Caspian. Merwan was aroused from his unaccountable slumber; but the star of his fortune was on the decline. On the banks of the Zab, between Mosul and Arbela, a spot rendered memorable by the final victory of Alexander over Darius, he met and encountered the forces of his antagonist. Here his troops were totally defeated. Overwhelmed with sorrow and disgrace, he wandered from the scene of conflict across the Euphrates, casting a melancholy look on his palace of Harran, and directing his steps towards the Syrian frontier. Everywhere the deserted monarch, —such is the lot of the unfortunate,—found the gates of his own cities closed against him. From Damascus he directed his flight through the valleys of Palestine to the borders of Egypt. Abdallah pursued the fugitive, tracking his route by the burnings and devastations which he had committed to retard the progress of the enemy. Followed up the Nile to Abousir, he was surprised in one of the Christian churches where he had taken refuge with a solitary attendant, and being transfixed with a lance, expired on the spot. While the pursuers gathered round to contemplate the spectacle of fallen greatness, a slave dismounted and put a final, perhaps a welcome, period to his misfortunes, by striking off his head. The dynasty of the Ommiades occupied the throne exactly eighty-nine years, though their power may be traced in a subordinate degree to the caliphate of Omar I., when Moawiyah succeeded to the government of Damascus. Historians have remarked with surprise, that while they maintained their ascendency under weak and dissolute princes, they should have found their extinction in the reign of a sovereign alike magnanimous in victory and defeat; and capable, by his splendid military talents, of restoring the ancient lustre of his family. Yet, with all these advantages on their side, and an army of 120,000 soldiers against a sixth part of that number, the white faction, from the Oxus to the Nile, was scattered with dismay before the sable legions of the Night and the Cloud.

Seldom is a change of dynasty accomplished without encountering the horrors and atrocities of revolutionary phrensy. The elevation of the house of Abbas would appear to have been attended with circumstances of sanguinary ferocity and deliberate cruelty that have rarely been surpassed in the annals of the East. The most distant branches of the hostile race were sought out with merciless industry, and cut off with indiscriminate revenge. At Damascus, fourscore of their chiefs were insidiously invited to a public entertainment; and the laws of hospitality were violated by a promiscuous massacre. Their mutilated carcasses were laid one upon another, forming a kind of platform, covered over with a slight carpeting, on which was spread the festive board; and where, with a barbarity truly savage, the governor (Abdallah) and his friends seated themselves to partake of a sumptuous repast, and celebrate the triumph of his party. When living victims were wanting, he proceeded to commit violence on the repositories of the dead. The sepulchres of all the caliphs and princes of the Ommiades, that of Omar II. alone excepted, were broken open, their contents burnt to ashes, and dispersed to the winds.

Of Abu Moslem, the author of the *Call* of the Abbassides, as he has been styled, we may here observe, that he experienced at last the usual ingratitude of despotic courts. His eminent services had obtained for him the lieutenancy of Khorasan, and the title of the Maker of Kings; but his arrogance, and certain suspicions against his loyalty, drew down upon him the resentment of the caliph, who ordered him to be privately assassinated. In the pilgrimage to Mecca, he had appeared with a rivalry of splendour that far outshone his sovereign. For the conveyance of his kitchen-equipage alone 200 camels were necessary. Even on ordinary occasions his munificence was extravagant. Twelve hundred mules or camels were requisite for his household baggage; he employed 1000 cooks, and the daily consumption of his table amounted to 3000 cakes, 1000 sheep, besides oxen, poultry, and other provisions. The steed and the saddle which had carried any of his wives were instantly destroyed and burnt, lest they should be afterward used by an individual of the other sex.

No warrior was more prodigal of life; and such was the stern ferocity of his temper, that he was never seen to smile except on the day of battle, and amid the horrors of blood and slaughter. He could boast with pleasure, perhaps with truth, that he had destroyed 600,000 of his enemies, exclusive of those who perished on the field. His eventful story, as we learn from Price, has been wrought into a very interesting romance, entitled the Abu Moslem Nameh, well known in the East, and in which, amid a tissue of extravagant adventure, many surprising facts have doubtless been interwoven.

CHAPTER X.

CONQUEST OF AFRICA AND SPAIN.

Renewal of the War in Africa—Victories of Akbah—Founding of Cairoan—Revolt of the Africans—Reduction of Carthage—Defeat of the Saracens by the Moors—Success of Musa—Expulsion of the Greeks and final Subjugation of Africa—Invasion of Spain by Tarik—Defeat of the Goths—Rapid Conquests of the Moslems—Surrender of Cordova, Seville, and Toledo—Reduction of the Country as far as the Bay of Biscay—Recovery of Seville—Descent into Languedoc—Preparations of the Saracens to subdue Europe—Recall and Disgrace of Musa—Arabian Settlers in Spain—Progress of the Saracens in France—Defeat of Eudes—Victory by Charles Martel at Tours—Expulsion of the Saracens from France—Success of the Moslems in the East—Sieges of Constantinople—Repulse of the Arabs—Their Conquests beyond the Oxus—Surrender of Samarcand—Invasion of India—Extent of the Mohammedan Empire.

Unwilling to blend the foreign conquests of the Saracens with their civil dissensions, or connect their military exploits with the history of individual caliphs, we have purposely left unnoticed the progress of their arms abroad after the death of Othman. Down to that period the Arabs had applied themselves solely to the extension of their dominions and the propagation of their creed. With the accession of Ali sprang those calamitous schisms which lighted up the fires of revolt in every province of their empire. Yet such were the energies of this surprising people, that while Syria, Persia, and Arabia were convulsed with rebellion, their cities laid in ashes, and their rivers died with mutual slaughter, their armies were extending the dominion of the Koran, and spreading their triumphs to the remotest regions of the East and the West. When the Ommiades had fixed themselves with hereditary order on the throne, and cemented the irregular fabric of their authority with the best and holiest blood of Arabia, the campaigns of plunder and proselytism were renewed; fresh squadrons issued from the Desert to tread in the footsteps and rival the glory of their predecessors.

In Africa the victories of Abdallah had been attended with no decisive result; the troops returned with their spoil to Medina, and it was not till after a lapse of twenty years (A. D. 648–668) that the cries of the oppressed inhabitants recalled them to that quarter. The Greek emperor was not ignorant of the tribute which the Arabs had exacted from his African subjects; but, instead of pitying or relieving their distresses, he imposed, by way of fine, a second tax of an equal amount. In the provocation of despair, they abjured the religion of the Roman government, and declared their preference of a Mohammedan to a Christian tyranny. The caliph, Moawiyah, equipped an expedition of 10,000 men, commanded by Bashar and another general of his own name, who penetrated as far as the territory of Carthage, took several important towns, defeated a force of 30,000 Greeks, and carried away 80,000 captives, together with immense spoil.

But the title of conqueror of Africa more justly belongs to Akbah, the succeeding lieutenant, and one of the governors of Egypt. He left Damascus at the head of a brave though not numerous army, and pushed his victories far into the interior. Of his Numidian conquest we have no certain account; for little credit is due to the oriental writers, who have peopled those regions with fictitious multitudes, and planted them with imaginary citadels and towns to the extravagant number of 360. Towards the seacoast his progress was defined by the well-known cities of Bujia and Tangier, the latter of which the

Arabian fables have decorated with walls of brass, and roofs covered with gold and silver,—emblematic expressions, perhaps, of its strength and opulence. Crossing the Atlas range and the Great Desert, the fearless Akbar traversed the wilderness in which the Moslems afterward erected the splendid capitals of Fez and Morocco. His career, though not his zeal, was checked by the prospect of a boundless ocean. Spurring his horse into the waves, and raising his eyes to heaven, he exclaimed, with the enthusiasm of the Macedonian madman, "Great God! if my course were not stopped by this sea, I would still go on to the unknown kingdoms of the West, preaching the unity of thy holy name, and putting to the sword the rebellious nations who worship any other gods but thee!"

It was on the banks of the Sus, which falls into the Atlantic not far from the Canary Islands, that he encountered the last of the Moors, a race of lawless undisciplined savages, who beheld with astonishment the strange and resistless banners of a foreign invader. They possessed neither gold nor silver to reward the conqueror; but a valuable traffic was discovered in the beauty of the female captives, some of whom brought in the Eastern markets 1000 pieces of gold. Numbers of the barbarians had professed the faith and joined the ranks of the Moslems; though their aid and their conversion were alike insincere. To curb their seditious spirit, as well as to afford a place of security for the Saracen troops and the immense booty they had amassed, Akbah adopted the judicious policy of founding an Arabian colony in the heart of Africa. In the fiftieth year of the Hejira, he laid the foundation of Kairwan or Cairoan, about fifty miles to the south of Tunis, and twelve westward from the sea. A circumference of 3600 paces was encompassed with a brick wall; and in the course of five years, from being the station of a garrison, it became the capital of the province and the residence of the governors. In later ages it increased in wealth and population, maintained a flourishing trade, and was celebrated for its stately buildings and its eminence as a seat of learning.

With all his precautions Akbah was unable to maintain his recent conquests. The crowds of discontented Greeks and Moors that had been attracted to his standard perfidiously threw off their fickle allegiance. This general defection recalled the impetuous conqueror from the shores of the Atlantic. Of the mutiny just ready to explode he was apprized by an Arab chief, who had disputed with him the command; and was then suffering in irons the punishment of his unsuccessful ambition. The insurgents had trusted to the discontent and revenge of the captive; but he generously disdained their offer, and preferred to die with his countrymen. The rebellious multitude surrounded them in their camp, and left them no resource but that of an honourable death. In the hour of danger, Akbah unlocked the prisoner's fetters and advised him to retire; but he chose to perish with his rival. The ardour of friendship revived; they unsheathed their scimitars, broke their scabbards, and maintained with the zeal of martyrs an obstinate combat, till they fell by each other's side among the last of their slaughtered band. Zobeir, the third commander of Africa, revenged on the natives the fate of his predecessor. He vanquished them in many battles, but his victories were not productive of any solid conquest; and in attempting to reduce the Greek dependencies on the seacoast, he was totally defeated by the powerful armament which the Emperor Justinian II. had sent to the relief of Carthage. Abdolmalec, on his accession, undertook the subjugation of the West on a sounder plan, and with far superior resources. The standard was intrusted to Hassan, governor of

Egypt; while the revenue of that wealthy province with 40,000 brave Arabs were devoted to that important service.

Hitherto the scimitar appears to have been the main implement in the wars of the Saracens. There was little room for military tactics, or the talents of the engineer, such as we find in the warlike operations of the Greeks and Romans. To lie in ambush, to surprise by assault, to invest a city by armies rather than by lines or walls, and patiently to await the effects of discontent or famine, were the only arts which the Moslems had employed in conquering the strong-holds of their enemies. But a more extended intercourse with nations better skilled than themselves in the mechanism of warfare taught them the use and advantages of more powerful instruments than the sword in conducting regular sieges. Hassan carried in his train a number of warlike engines. The interior of the country had been overrun by his predecessor; but the Mediterranean coast still remained in the hands of the Greeks. Carthage, the proud capital of Africa, had been yet unassailed, and the number of its defenders was recruited by the fugitives from Cabes and Tripoli. In the year 697 it was reduced and pillaged, in spite of the succours from Justinian II., and from Egiza, the Gothic king of Spain, who was anxious to repel the tide of Saracen conquest before it should rush fiercely on his own shores.

But their triumph was soon disturbed. The Christians, under John the Patrician, a general of great experience and renown, embarked at Constantinople for the seat of war. The expedition was joined by the ships and soldiers of Sicily, and a formidable reinforcement of Goths from Spain. The confederate army appeared off Carthage, and breaking the chain that guarded the entrance to the harbour, they effected a landing. The citizens hailed the ensigns of the Cross as the emblems of victory and deliverance. The Arabs were compelled to evacuate the town, and retire to their camps in the desert; but in the course of the following spring they equipped a new and more numerous armament by sea and land, when the Christian patrician in his turn was driven from the capital, which was now consigned to the flames. In the vicinity of Utica a second battle was fought: the Greeks and Goths were defeated; and to save themselves from total extermination, they set sail for Constantinople. But their expulsion did not leave the Arabs entirely masters of the country. In the inland parts the Moors, or Berbers, a name anciently applied to all nations except the Greeks, and finally restricted to the inhabitants of a local district on the northern coast of Africa, maintained a disorderly resistance to the arms and the religion of the Saracens. The fame of Cahina, whom they acknowledged for their queen, and even revered as a prophetess, attracted the roving tribes to her standard. Her enthusiasm inspired them with union and energy. Hassan gave battle; but his veteran bands were repulsed by the superior numbers of the Moors. A single day lost the conquests of an age, and the Moslem chief retired to the confines of Egypt, where he waited five years the expected succours of the caliph.

When the invader had gone, Cahina assembled her followers, and recommended a mode of defence extraordinary in itself, but not unusual or inconsistent with the barbarous policy of wandering hordes. "Our cities," said she, "and the gold and silver which they contain, perpetually attract the arms of the Arabs. These vile metals are not the objects of our ambition: we content ourselves with the simple fruits of the earth. Let us destroy those cities; let us bury in their ruins those pernicious treasures, and when the avarice of our foes shall be destitute of temptation, perhaps they will cease to disturb the tranquillity of a warlike people." The advice of Cahina was adopted; but the sight of their desolated country made the Moors repent of their frantic passion. They bewailed the devastation produced by their own hands, and began to prefer the dominion of the Saracens to the yoke of the Byzantine emperors, or even the disorderly rule of their own sovereigns.

Hassan returned, but he met with less determined opposition. Revolt weakened the enemy; in the first battle the royal prophetess was slain, and her death overturned to its basis the insecure fabric of her superstition and her empire. Under the successor of Hassan the spirit of insubordination again incited these turbulent savages to acts of open rebellion; but it was finally quelled by the activity of Musa and his two sons, who crowned a decisive victory by carrying away 300,000 captives, the fifth of whom, belonging to the caliph, were sold for the benefit of the public treasury. Musa was more than a conqueror; and the success which he had acquired as a general was confirmed by his talents as a preacher and a statesman. He used the utmost diligence in diffusing the doctrines of the Koran and the rites of the Mohammedan faith.

The Arabian writers are so vague and incorrect in their geographical statements, that it is scarcely possible to trace the limits of Mussulman dominion in the heart of Africa. They appear, however, to have extended over the whole maritime coast, and as far inland as the verge of the Great Desert. A small portion of Mauritania was still wanting to complete their western conquests. Tangier, the capital of that province, was besieged and taken by Tarik, the servant of Musa, to whom he had intrusted the van of his army. The general himself laid siege to Ceuta, a fortress in possession of the kings of Spain. Here, by the wisdom and bravery of the governor, Count Julian, a Christian, he failed in his attempt. The place, however, was soon after carried by the treachery of the count himself, who abandoned his rightful sovereign and made a tender of his person and his sword to the Arabian leader.

A desultory war of forty years had thus sufficed to drive the Greeks from their African possessions. The natives were solemnly invited to renounce the faith of the Cross, and accept that of Mohammed. For a time they preferred to purchase freedom of conscience and religious worship by the payment of an annual tax. By degrees their reluctance was overcome; their minds were tempted by the invisible as well as the temporal blessings of the Arabian prophet. They had but to repeat a sentence and submit to the rite of excision, and the subject or the slave, the captive or the criminal, were from that moment free, honoured as companions, and invested with the prerogatives of citizens. Motives of interest or convenience might yield to serious convictions, and a race of sincere proselytes would spring up with the rising generation. So rapid was the progress of the Koran, that in less than fifty years after the conquest effected by Musa, the lieutenant of the first Abbassidan caliph could inform his master that the tribute imposed on the Mogrebin or western infidels, then under his government, had totally ceased through their unanimous adoption of the true faith.

In their climate and government, as well as in their diet and manners, the inhabitants of Africa resembled the wandering Bedouins of the Desert; it was therefore no difficult enterprise to accustom them to believe in the apostle of God, and obey the commander of the faithful. With the religion they were proud to adopt the name and the language of Arabs. The blood of the strangers gradually mingled with that of the vanquished natives. Thirty thousand of the barbarian youth were enlisted in the battalions of Walid; while colonies of Arabian emigrants, abandoning their own country, crossed the Nile and scattered their tents over the sandy plains of Libya. Though a few of the Moorish tribes still retained their peculiar idiom, with the appellation and character of White Africans, yet such has been the influence of time and intercourse in softening down national distinctions, that one and the same people seem to have diffused themselves over the vast regions between the Euphrates and the Atlantic.

During the fifth century Africa had been the theatre of religious war. The hostile fury of Moors, Vandals, and Donatists had overturned 500 episcopal churches; the people languished without discipline or knowledge. The doctrines of Cyprian and Augustine were no longer studied; Christianity itself, driven before the tide of invasion, finally abandoned the southern coasts of the Mediterranean; the only land in which the light of the Gospel, after a long and perfect establishment, has been totally extinguished.

From Africa the transition to Spain was easy, nor was it altogether untried; for the Moslems, under Abdallah, had visited the shores of Andalusia as early as the time of Othman. This country, after witnessing the triumphs, and becoming tributary to the power of the Carthaginians and Romans in succession, had submitted, early in the fifth century, to the government of the Goths, the most formidable of the northern invaders. But these impetuous conquerors no longer resembled the fierce soldiers of Alaric, who had marched victorious over the wide dominions of the Cesars, from the borders of Scandinavia to those of the Atlantic. Without divesting themselves of their primitive rudeness, they had adopted all the false refinements of the vanquished nations, and passed by rapid steps from the extreme of ignorance and poverty to that of luxury and vice.

After the decease or deposition of Witiza, his two

sons were supplanted by the ambition and the intrigues of Roderick, the last of his race, whose father had been a provincial governor. But the materials of revolution were lurking in the bosom of the country, and the smallest spark was sufficient to throw them into combustion. Count Julian, by his rash invitation of a foreign power, was the individual that fired the train.

The influence of this nobleman rendered him a useful subject or a formidable enemy. His estates were ample; his followers bold and numerous; and as governor of Andalusia and the opposite province of Mauritania, it was evident he held in his hand the keys of Spain. The alleged seduction of his daughter, Cava, by Roderick, made him a rebel and a traitor to his king. Too feeble, however, to venture with his own resources on the execution of revenge, he determined to implore the aid of the Saracens; and crossing the sea, he repaired to the camp of Musa. In a personal interview with that general, he revealed the weakness and the wealth of his country; for in the abundance of its gold and silver Spain was the Mexico and Peru of antiquity.

Before embarking in this new conquest, Musa sent to obtain the permission of Walid. The answer of the caliph was favourable; but it implied that the science of geography had been little studied at the court of Damascus. He cautioned him not to venture rashly with the Moslems on the navigation of a perilous ocean; and directed him to make an incursion into the country, that he might previously ascertain its actual state. The lieutenant, in reply, gave the emperor of the faithful to understand, that the sea between Africa and Spain was not a tremendous ocean, but merely a narrow strait (Sebtah) which the eye could reach across; in compliance with his instructions, however, he resolved to make a previous trial of this unknown region. At once to prove the sincerity of Julian, and obey the cautious policy of Walid, he ordered the count to make the first hostile experiment himself; and accordingly, with a body of troops collected from his own government, he made a predatory descent (July, 710) with two ships on the coast of the Verdant Island, for so the Arabs termed the opposite shore at Algesiras, near which stood the town and castle of Julian. Musa despatched a second expedition of 500 troops, under one of his officers, Tarifa, who effected a landing at a spot which still bears his name; and penetrating into the country, they carried off much plunder, among which was a female captive, more beautiful than any the Saracens had yet beheld.

In his residence at Tangier, Musa continued with success to hasten his preparations; and in the ensuing spring (A. D. 711), 7000 men were embarked, under the command of Tarik, already distinguished as a brave and skilful soldier. The place where they landed was at Mount Calpe, one of the Pillars of Hercules; and in its modern appellation of Gibraltar (Gebel al Tarik, or Hill of Tarik) the name of the hero is still preserved. Here he formed his first camp, the intrenchments of which were the original outline of those fortifications that have rendered this singular rock so important as a military station in the hands of Britain. By getting possession of Algesiras, the port of Andalusia, Tarik opened a passage into the country, which he subdued as far as Cadiz.

When intelligence of the descent and progress of the Saracens was conveyed to Roderick, he was engaged in a war against the insurgent Bascons, in the district of Pampeluna. The defeat of his lieutenant, Edeco, whom he had haughtily commanded to seize and bind the presumptuous strangers, admonished him that the danger was imminent, and must be averted without delay. Hastening to Cordova, he took up his residence in the castle of that place. In a short time the king of the Romans, for such is the title the Arabs gave the Gothic monarch, saw himself at the head of 90,000 or 100,000 men,—a formidable power, had their fidelity and discipline been equal to their numbers. Tarik, on learning the superiority of the enemy, applied for assistance to Musa, who was actively employed in collecting troops and preparing transports. A reinforcement was instantly despatched; and a body of 12,000 Moslems, under the conduct of Julian, who undertook to guide them through the passes of the hills and gather information, advanced to the neighbourhood of Cadiz, eager for pillage, and anxious to try the metal of their scimitars against the terrible subverters of the Western Empire.

Roderick advanced to meet the foe, and the two camps were divided by the small stream of the Guadalete. The scene of the memorable battle that determined the fate of Spain is generally understood in Europe to have been near the town of Xeres; though the Arabs, if their geography can be trusted, place it at Medina Sidonia, south from Cadiz. For seven days, from Sunday till Sunday, the Goths and the Saracens encountered each other's strength in skirmishes and single combats. But the influence of Julian, and the discontent of the Christian chiefs, were secretly spreading the leaven of defection among the ranks of Roderick. The two sons of Witiza, who commanded the right and left wings of the Gothic army, had, in the hope of regaining their father's throne, stipulated with Tarik, previously to the engagement, to desert the usurper in the midst of the battle, on condition that the Arabian general, if victorious, should secure to them their patrimonial inheritance in Spain amounting to 3000 valuable farms or manors. The latter did not hesitate to accept their proposal, and only waited an opportunity to turn their defection to the best account.

It was not till the eighth day (about the 25th of July) that the two armies joined in the deadly and decisive conflict. Roderick, who had brought with him to the field a splendid retinue of wagons containing his treasures, was dressed, or rather cumbered, with a flowing robe of gold and silken embroidery; a diadem of brilliant pearls adorned his head, over which was expanded a canopy set with rubies and emeralds; and his throne, a litter or couch of ivory on which he reclined, was borne between two white mules. In this equipage, much more suited to the luxury of an Asiatic court than a campaign, he appeared before his troops, and harangued them on the importance of the objects for which they were contending. Tarik sustained the valour of his fainting companions by appealing to the recollection of their former exploits. "My friends," continued he, in the brief but touching eloquence of his country, "the enemy is before you,—the sea is behind! Whither would you fly? Follow your general! I am resolved either to lose my life, or trample on the prostrate king of the Goths!"

Both sides maintained the bloody combat with their characteristic fury; until the well-timed desertion of Oppas, archbishop of Toledo, and the two princes his nephews, turned the tide of battle in favour of the invaders. The two wings had given way, yet for a while Roderick maintained his ground with the centre, vainly endeavouring to recall his dispersed and terrified squadrons. His own courage at length forsook him; he started from his gaudy palanquin, mounted Orelia, the fleetest of his steeds, and, like the rest of his nobles, consulted his personal safety amid the general disorder.

The genius of Spain has contrived to throw a veil of romance over the fate of Roderick after his departure from the plain of Xeres. Some ridiculously assert that he escaped and took refuge in a hermit's cell; and Cervantes, in his inimitable fiction of the chivalrous Quixote, has cast him alive into a tub full of serpents, which are made to inflict on his body the peculiar penance which his crimes deserved. The Arabian writers tell, what is probably the truth, that though he fled, he only avoided a soldier's death to perish more ignobly in the waters of the Bœtis (or Guadalquiver). The Moslems found his diadem and robes cast on the bank; and his horse, bearing a saddle covered with gold and rubies, plunging in the mud, where one of his boots was also discovered sticking; leaving no doubt as to the fate of the vanquished prince;—" an end," as a valiant historian of the Arabs remarks, " worthy of those kings who withdraw from the field of battle." As the body was lost in the stream, some meaner head must have been exposed in triumph at Damascus to gratify the pride and ignorance of the caliph.

This victory the Saracens purchased at the expense of 16,000 lives. The field was strewn with their slain; yet the Goths suffered more severely. A pursuit of three days scattered or destroyed the remains of their army; their chiefs and nobles who had fallen were distinguished by the rings of gold on their fingers; those of inferior condition by trinkets of silver; and the slaves by similar ornaments of brass.

The news of Tarik's success, and the spoils that accompanied it, were no sooner conveyed across the straits than crowds of adventurers flocked to him from all quarters, passing the narrow sea in every boat or bark they could find. The sight of this fresh invasion obliged the frightened Spaniards to quit the coast and the plain country, and betake themselves to their mountains and fortresses. The conqueror, by the advice of Julian, who had now plunged too deeply into guilt to expect reconciliation, adopted measures to seize the capital, Toledo, without delay, and subdue the entire country, before the distracted inhabitants had time to elect a new sovereign.

Detachments were despatched to reduce the most important of the provincial towns. A body of 700 horsemen, mounted on the animals taken from their slaughtered enemies, assaulted Cordova. The principal inhabitants had fled; but the common people and the commander of the city remained, with a garrison of 400 men. Favoured by the darkness of the night, and a convenient shower of hail, which drowned the tread of the cavalry, the besiegers scaled the ramparts, killed the guards and took possession of the town. The governor and his troops posted themselves in a solitary church, where, being supplied with water conveyed under ground, they maintained a resistance of three months. But the spring that supplied them was discovered by a slave, and stopped; and on their obstinately refusing the usual conditions, the church was burnt, most of the Christians perishing in the flames. Malaga, Granada, and the coast as far as Murcia yielded in rapid succession to the forces of the invaders; nor could the ingenuity of Tadmir or Theodomir, the Gothic prince, prolong the independence of his capital, Orihuela, by parading his women on the walls in the dress and arms of soldiers, to conceal his weakness and deceive the enemy.

The march of Tarik was attended with equal success; most of the towns surrendered or were taken by force. Carmona fell,—the people of Seville consented to pay tribute,—Ecija stood a siege; but after an obstinate battle, where many of the Moslems were killed or wounded, peace was granted on the usual terms. His progress from the Bœtis to the Tagus met with little interruption. The inhabitants had abandoned the open country, and to increase their terror, Tarik caused his men to cook the flesh of the slain in presence of the captives, some of whom were allowed to escape, that they might spread the astounding report, as if their ferocious invaders delighted, not only in shedding blood, but in feasting like cannibals on the mangled limbs of their slaughtered foes.

Directing his steps through the Sierra Morena that divides Andalusia from Castile, the Saracen general appeared with his victorious band under the walls of Toledo. The Catholic part of the inhabitants had fled to a dependent town beyond the mountains, carrying with them the relics of their saints. The Jews and others that remained were glad to surrender on a fair and reasonable capitulation. The voluntary exiles were permitted to depart with their effects. The Christians were allowed seven churches for the use of their worship; the archbishop and his clergy, the monks and the magistrates, were left free to exercise their respective functions. The Hebrews, who were received and trusted by the conquerors in preference to the Christians, and to whose secret or open aid Tarik was deeply indebted for his success, were treated with kindness and generosity. Nor were they ungrateful to the restorers of their political liberties; for the friendship between the disciples of Moses and Mohammed was maintained till the final era of their common expulsion. Among the valuable plunder of Toledo are enumerated 170 crowns formed of pearls, rubies, and other precious stones. It is also related, that there was a gallery or hall full of gold and silver vessels, so large as to accommodate a body of horsemen in their diversion of throwing the spear.

From the Gothic metropolis, Tarik, pursuing the fugitives, spread his conquests to the north over the realms of Castile and Leon. Passing the Asturian mountains, he bounded his victories by the maritime town of Gijon; and westward, by the city of Astorga. With the speed of a traveller, he had extended his march 700 miles, from the Rock of Gibraltar to the Bay of Biscay; and, in the course of a few months reduced a country which, in a more savage and disorderly state, had resisted for 200 years the arms of the Romans.

The rapid conquests of his lieutenant had at first gained the applause of Musa; but his increasing renown created a spirit of envy. Jealous of his own fame, he began to apprehend that Tarik would leave him nothing to subdue. Committing the Moorish provinces to the care of his eldest son, Abdallah, he crossed over to Spain at the head of 10,000 Arabs and 8000 Africans. Seville had thrown off its allegiance to the Moslems; but, after a siege of some months, it surrendered to Musa. Hence he passed on to Merida, a strong city, formerly the seat of the Lusitanian government. When the Arab chief beheld the aqueducts, theatres, temples, and other works of Roman magnificence, "I should imagine," said he, "that the human race must have united their art and power in the founding of this city. Happy is the man who shall become its master." The inhabitants repulsed the invaders with determined bravery. Disdaining the confinement of their walls, they sallied forth and gave battle on the plain; but their return was intercepted, and their temerity chastised by an ambuscade. The besiegers rolled their wooden turret forward to the foot of the rampart; but the citizens assailed them so furiously, that they were compelled to retire; and the Tower of the Martyrs long commemorated the fall of those who had been slaughtered in this machine.

An artifice of Musa is said to have hastened the terms of capitulation. At his first interview with the deputies of the place, his hair and beard were undressed, and white as wool. Next day they were surprised to find his beard red; and on the third still more astonished to find it of a black colour. Being totally unacquainted with the Arabian custom of staining the hair, they represented to their fellow-townsmen the hopelessness of resisting a general who was a prophet, who could change his appearance at pleasure, and even transform himself from age to youth. The conditions were accepted; the inhabitants saved their own properties, but consented that the effects of the fugitives and the slain, with all the riches and ornaments of their churches, should be delivered up to the Saracens.

Musa advanced towards Toledo; but Tarik, informed of his approach and aware of his intentions, went forth with a retinue of his chiefs to receive him; and the two conquerors met near Talavera. Their first salutation was cold and formal. Proceeding to the capital, they entered together the palace of the Gothic kings. Musa exacted a rigid account of the booty and public treasures in his hands. He even carried his animosity so far, if we may believe Cardonne, as to strike him with his whip, to load him with abuse, and even with chains: all of which indignities the patient hero bore; so pure was the zeal, and so high the sense of discipline and subordination in the Arabian armies. The merit and probity of Tarik appear to have convinced him that his suspicions and his resentment were alike unjust. He gave him the hand of friendship, and restored him to the command of the van.

From Toledo the Moslems continued their victorious march northward. Saragossa was reduced, and a mosque erected by the liberality of the Koreish. The port of Barcelona was opened to the vessels of Syria. The whole provinces as far as the ocean and the Pyrenees submitted without resistance; such was the alarm which the Saracen name had inspired. The Goths were chased beyond the mountains into the territory of Languedoc; and a column or statue, erected by Musa at Carcassone, marked the limit of his progress in Narbonnese Gaul. In the church of St. Mary, at this latter place, were seven equestrian statues of massy silver, which the conqueror probably carried off as the first trophies he had won on the soil of France. Tarik, after taking the city of Narbonne, and one or two obscure fortresses, marched onwards as far as the

Rhone; but the preparations of the king of the Franks who had taken alarm at this irruption of the barbarians, compelled them to retreat.

During the absence of Musa in the north, his son Abdolaziz was occupied in confirming or extending their acquisitions in other parts of Spain. He reduced the remainder of the Mediterranean coast from Malaga to Valencia; obliged the governor, Prince Theodomir, to deliver up his seven cities; bound him neither to assist nor form alliance with the enemies of the caliph; and to pay annually for himself and each of his nobles one piece of gold, four measures of wheat, as many of barley, with a certain proportion of honey, oil, and vinegar; each of their vassals being taxed at one moiety of the said impost. On these conditions the Goth was to continue undisturbed in his principality.

The whole peninsula, one solitary corner excepted, being reduced, Musa formed the bold design of making himself master of all Europe. With a vast armament, by sea and land, he was preparing to repass the Pyrenees, to subvert the kingdom of the Franks in Gaul, then distracted by the wars of two contending dynasties; to extinguish the power of the Lombards in Italy, and place an Arabian imam in the chair of St. Peter. Thence, after subduing the barbarous hordes of Germany, he proposed to follow the course of the Danube, from its source to the Euxine Sea, where he would have joined his countrymen under the walls of Constantinople.*

* The conquests of Spain and Africa are passed over silently or slightly by Abulfeda, Abulfarage, Elmacin, Tabiri, and the other oriental annalists; but the chasm is supplied by Cardonne (Hist. de l'Afrique et de l'Espagne sous la Domin. des Arab.), Leo Africanus and Marmol (Descrip. de l'Afrique), Chenier (Recherch. Hist. sur les Maures), Mariana (Hist. de Reb. Hisp.), Casiri, who has collected many fragments of Arabian literature (Biblioth. Arabico-Hisp.), Roderick of Toledo (Hist. Arab.).

These daring projects, however extravagant they may appear, admitted of easy execution, and might perhaps have been realized, had not the scheme, by some channel, been communicated in an unfavourable light to the caliph, who was greatly terrified at the risk his forces would incur in such an enterprise, and despatched an envoy to recall Musa, and in case of his refusal to bring the Moslems back himself. The adventurous hero was engaged in subduing the Galicians, and so intent was he on this invasion, that he bribed the forbearance of the messenger, by offering him half of his own share of the spoils. This delay was followed by a harsher and more peremptory summons. A second envoy from Syria entered the camp of Musa at Lugo, and arrested the bridle of his horse in presence of the whole army. His own loyalty, or that of his troops, suggested the duty of instant compliance. Leaving his two governments in the hands of his two sons, Abdallah and Abdolaziz, he set out for Damascus, to answer in presence of the caliph, not only for disobedience of orders, but for his proceedings in regard to Tarik, whose friends had conveyed to the royal ear a true statement of the services and wrongs of that distinguished chief. Musa and his injured rival left Cairoan together, preceded by a vast booty, which displayed in long triumph the spoils of Africa and Spain. Four hundred of their nobles, with golden coronets and girdles, attended his train; besides 30,000 female captives, selected for their birth or their beauty.

On reaching Egypt he was informed of the dangerous illness of Walid, by a private message from Soliman, the presumptive heir, who requested him to halt, wishing to reserve for himself so splendid a prize. Musa proceeded; but on arriving at Damascus he found an enemy on the throne. In his trial before a partial tribunal, he was convicted of avarice and falsehood, and amerced for his rapacious exactions in a fine of 200,000 pieces of gold. His accusers charged him with concealing a jewel more valuable than any that had been seen since the conquest of Persia. The justice of Tarik's cause was acknowledged, and his unworthy treatment amply revenged.

Among the contested spoils was the celebrated Table said to have belonged to Solomon, and brought from Jerusalem to Rome, where it was found by the Goths when they sacked and burnt that ancient capital. Whatever may have been its origin, whether a Jewish or a Christian relic (for the latter were in the custom of bequeathing such valuable property to their churches), its richness and workmanship are highly extolled. The fabric was of pure gold, others say one solid piece of green jasper, set with the most precious rubies and emeralds. Its feet, composed of the same materials, are reported to have amounted to 365 in number. This famous trophy was found in the palace or cathedral of Toledo when it surrendered to Tarik, and constituted one of the charges brought by him against Musa, who claimed it for himself, and denied it had ever been in the possession of his rival. But this assertion was disproved on the spot; for the wily Tarik had taken the precaution to break off one of the legs, which he now produced from under his dress, and convinced the court that the splendid booty had been first in his hands. New indignities were heaped on the degraded commander. After a public whipping, he was compelled to stand a whole day in the sun before the gate of the caliph. So utterly was he reduced by heavy exactions, that on being released from prison he was led about like another Belisarius, as a common beggar, to solicit from public charity the scanty means of subsistence. Nor was the resentment of Soliman appeased by the ruin of the father: his vengeance demanded the extirpation of the whole family. A sentence of death was despatched with secrecy; and the two governors of Africa and Spain were its victims. Abdolaziz, who had married Egilona, the haughty widow of Roderick, was slain in the mosque or palace of Cordova. By a refinement of cruelty, the head was brought to Mecca, where Musa lingered in wretchedness and exile; "Know ye the features of a rebel?" said the taunting messenger of the caliph. "I know my son," said the aged chief with indignation; "I assert his innocence; and I imprecate a juster fate against the authors of his death." This truly great and successful commander expired of grief and absolute want at his native place in Hejaz (A. H. 92). His fate reflects disgrace on his ungrateful sovereign, though it might serve to convince the astonished auxiliaries of the Arabian armies in the West, that services however meritorious, or crimes however scandalous, could never escape the potent jurisdiction of the caliph.

The Spanish revolution, so speedily effected, was marked by the characteristic barbarity of the invaders. The licentious soldiery pillaged the towns, profaned the churches, and desolated the country: while the miseries of the vanquished, as a native historian has remarked, appeared to constitute the happiness of the victorious general. The disciples of the Koran availed themselves fully of its liberal principles in satisfying their passions of avarice and voluptuousness; nor can their moderation be much applauded even when contrasted with the invasion of that peninsula by the Goths, or its recovery from the Moslem yoke by Ferdinand of Castile.

Spain, like other conquered countries, gradually lost its nationality, became reconciled to the yoke, and assimilated to the habits of its victors. The tincture of Gothic, Roman, and Punic blood, which it had successively imbibed, in a few generations disappeared in the name and manners of the Arabs. A numerous train of civil and military followers attended the camps of the governors; crowds of adventurers from the East poured in, who preferred a distant fortune to indigence at home; and the different cities where they were established assumed the name of the particular tribe or country of the new colonies. Settlers from Damascus occupied Granada and Cordova; Seville, Jaen, Xeres, and Malaga were planted by emigrants from Emesa, Kinnisrin, Palestine, and the banks of the Jordan. The natives of Yemen and Persia were scattered around Toledo and the inland country; Murcia and Lisbon the mingled hosts of Tarik and Musa shared with their brethren from Egypt; while the fertile valleys of the south were bestowed on 10,000 horsemen of Syria and Irak, descendants of the purest and noblest of the Arabian tribes. Ten years after the conquest, a map of the kingdom was presented to the caliph, with a description of the seas, rivers, and harbours, the inhabitants and cities, the climate, the soil, and the mineral productions.

The victorious Moslems had already crossed the Pyrenees, and annexed to their acquisitions the whole provine of Languedoc, which belonged to the Gothic monarchy of Spain. The project of extending their arms northward was resumed, to which nothing could be more favourable than the corrupt and tottering state of the Frankish government. The first invasion of the Saracens was bravely repulsed by Eudes, duke of Aquitaine, who had assembled under his standard a numerous army of Goths, Grecians, and Franks. Zama, the lieutenant of the caliph, lost his army and his life under the walls of Toulouse. This disaster stimulated the ambition and the vengeance of his sovereign; and the famous Abdalrahman (or Abderame, as he is called by the French historians), whom the Caliph Hesham had restored to the wishes of the soldiers and the Spanish colonists, undertook another expedition (A. D. 731), with the daring resolution of adding to the faith of the Koran whatever yet remained unsubdued of France or of Europe Having suppressed the domestic insurrection of Munuza, a Moorish chief, who had accepted the alliance and the daughter of Eudes, Abdalrahman, at the head of a formidable host, traversed the Pyrenees, and hastened without delay to the passage of the Rhone and the siege of Arles. The Christians attempted the relief of the city; but they were routed with severe loss. Many thousands of their dead bodies were carried down the stream to the Mediterranean, and the tombs of their leaders were still visible in the thirteenth century.

Westward, the progress of Abdalrahman was not less successful. He passed, without opposition, the Garonne and the Dardogne; but he found beyond these rivers the intrepid Eudes, who had formed a second army. After a bold resistance the duke sustained another defeat, so very fatal to the Christians that, according to the confession of Isidore, bishop of Badajos, God alone could reckon the number of the slain. From Bordeaux the impetuous Saracens overran the provinces of Aquitaine, whose Gallic names are disguised rather than lost in the modern appellations of Périgord, Saintonge, and Poictou. Tours and Sens were compelled to open their gates to the conqueror, while detachments of his troops overspread the kingdom of Burgundy as far as the cities of Lyons and Besançon.

Everywhere the track of the invaders was marked with fire and sword; for Abdalrahman spared neither the country nor the inhabitants. The memory of these devastations (to use the rapid and glowing narrative of Gibbon) was long preserved by tradition; and the invasion of France by the Moors or Mohammedans affords the groundwork of those fables, which have been so wildly disfigured in the romances of chivalry, and so elegantly adorned by the Italian muse. In the decline of society and art the deserted cities could afford but a slender booty to the Moslems; their richest spoil was found in the churches and monasteries, which they stripped of their ornaments, and delivered to the flames.

A victorious line of march had been prolonged above 1000 miles, from the Rock of Gibraltar to the banks of the Loire. The repetition of an equal space would have carried the Saracens to the confines of Poland and the Highlands of Scotland. The Rhine is not more impassable than the Nile or the Euphrates, and the Arabian fleet might have sailed without a naval combat into the mouth of the Thames. The seven Saxon kingdoms of Britain, torn by wars and factions, must have presented but a feeble barrier to the Eastern invaders, whose hardy frames seemed equally adapted to all climes and all countries. The Heptarchy, which the victorious arms and judicious policy of Egbert had united, might have passed into the hands of a viceroy from the court of Bagdad. Perhaps the interpretation of the Koran might have become the scholastic divinity in the halls of Oxford and Edinburgh. Our cathedrals might have been supplanted by the gorgeous mosque; and our pulpits employed in demonstrating to a circumcised people the truth of the apostleship and revelations of Mohammed. Such was the destiny that seemed to impend over all Europe, from the Baltic to the Cyclades, when the standard of Islam floated over the walls of Tours.*

From these probable calamities was Christendom delivered by the genius and fortune of one man. Charles Martel was the illegitimate son of the elder Pepin, and enjoyed the title of Mayor or Duke of the Franks. In a laborious administration of twenty-

* Les Arabes vainqueurs auraient planté les éstandardes de l'Islamisme sur les rivages de la Baltique.—*De Marlé's Hist. de la Domin. des Arab. en Espagne*, tome i. p. 141. *Heeren, Essai sur l'Influence des Croisades*, p. 265. *Turner's Hist. of Eng.* vol. i. p. 314, and iv. p. 409. *Gibbon*, chap. 51

four years, he had restored and supported the dignity of the throne by crushing the rebels in Germany and Gaul. In the public danger the hopes of his country turned to this active and successful warrior, and he was summoned to the command in place of his rival the Duke of Aquitaine. "We have long heard," exclaimed the terrified Franks, "of the name and conquests of the Arabs. We were apprehensive of their attacks from the East; they have now conquered Spain, and invade our country on the side of the West. Yet their numbers and, since they have no bucklers, their arms are inferior to our own."—"If you follow my advice," said the prudent mayor of the palace, "you will not interrupt their march, nor precipitate your attack. They are like a torrent which it is dangerous to stem in its career. The thirst of riches and the consciousness of success redouble their valour, and valour is of more avail than arms or numbers. Be patient till they have loaded themselves with the encumbrance of wealth. This spoil will divide their councils and assure your victory." Having collected his forces, Charles sought and found the enemy in the centre of France, between Tours and Poictiers. His march was covered by a range of hills, and the Arabian general appears to have been surprised by his unexpected presence. The nations of Asia, Africa, and Europe advanced to the encounter with equal ardour. In the first six days of desultory combat, the cavalry and archers of the East maintained their advantage, but in the closer onset of the seventh day, the oriental ranks were oppressed by the strength and stature of the German auxiliaries, who, "with stout hearts and iron hands," vindicated from the grasp of despotism the civil and religious freedom of their posterity. Charles wielded a huge mace; and the epithet of Martel, or the Hammer, which he earned on this occasion, is expressive of the resistless force with which he dealt his blows. Abdalrahman fell;

and after a bloody field the Saracens retired in the close of the evening to their tents. In the confusion and despair of the night, the motley tribes of Yemen, Syria, Africa, and Spain were provoked to turn their arms against each other. The remains of their host were suddenly dissolved, and each leader consulted his safety by a hasty and separate retreat. At the dawn of day, the stillness of the enemy's camp was suspected by the victorious Christians; but the report of their spies dispelled these apprehensions, and they ventured to explore the riches of the deserted tents. The joyful tidings were soon diffused over the Catholic world, and formed a theme for exercising the fancies of the credulous. We are told of three consecrated and miraculous sponges, which rendered invulnerable the French soldiers among whom they had been shared; and the monks of Italy are made to affirm that no fewer than 375,000 of the Mohammedans had been crushed by the hammer of Charles, while only 1500 Christians were slain.* The victory of the Franks at Tours was complete and final. Aquitaine was recovered by the arms of Eudes. The Arabs never resumed the conquest of Gaul; and they were soon driven entirely beyond the Pyrenees by Pepin, the son of Charles, who in the year 759 dispossessed them of Languedoc, Provence, and other parts in the south of France.

* Gibbon sarcastically records this erroneous computation; but had he chosen to examine or cite the original French authorities, he might have discovered that the mistake of the Italian chroniclers, Paul Warnefrid and Anastasius, arose from their substituting the entire number of the Saracen army for that of the slain. The words of John de Montreuil are "385 millia Macometicos *in fugam consertit.*" The services of Charles Martel in rescuing Christendom were but indifferently rewarded at last, if we may believe the legendary annals of the times, which affirmed that his corpse was most miserably dragged out of the grave by wicked spirits, while his immortal part was consigned to damnation, because he had appropriated great part of the ne to pay his soldiers.—*Baron. Annal. Eccles.* A. D. 741

Having seen the progress of these conquerors arrested in the West, we must turn our eyes once more to their exploits in the East; for while Tarik and Abdalrahman were trying the edge of their scimitars against the Goths and Franks in Europe, another army was extending the terrors of the Saracen name into the frozen regions of Tartary, and a third occupied in reducing Asia Minor and the capital of the Greek emperor. So early as the reign of Moawiyah, the Arabs had appeared under the walls of Constantinople. Yezzid the son of that prince, commanded the expedition. The troops suffered the extremity of hardship on their march; but their courage was animated by a genuine or fictitious saying of the Prophet, that to the first army who should take the city of the Cesars their sins would be forgiven. During many days the line of assault was extended from the Golden Gate to the eastern promontory, and the foremost ranks were impelled by the weight and enthusiasm of the succeeding columns. But the besiegers had miscalculated the strength and resources of the imperial metropolis.

Baffled by this firm and effectual resistance, the Arabs turned their arms to the more congenial occupation of plundering the European and Asiatic coasts of the Propontis. From April to September their piracies were continued; but on the approach of winter they retreated to the isle of Cyzicus, where they had established their magazine of spoil and provisions. During the six following summers, such was their enthusiasm and perseverance, the same mode of attack and retreat was repeated, with a gradual abatement of hope and vigour, till the accumulated mischances of shipwreck and disease, of fire and sword, compelled them to relinquish the fruitless enterprise. Thirty thousand martyrs had fallen during the siege of Constantinople, and among these was Abu Ayub, one of the Ansars and last of the companions of Mohammed, who had fought by his side at the battles of Bedr and Ohud. The mosque bearing his name rose on the tomb; and so much was its sanctity esteemed by the Turks, that they selected it for the inauguration of their sultans, who are here girded with the sword of state on their accession to the throne.

By this repulse the glory of the Saracen arms suffered a momentary eclipse. Constantine assumed the attitude of a conqueror. His ambassador appeared and was favourably received at the court of Damascus; in a general council a truce of thirty years was ratified between the two contending nations; and for the first time in the history of the Moslems, we behold the singular occurrence of their paying tribute to the Greek emperor, at a time when they were in possession of his fairest territories in Asia. The annual tax imposed on the commander of the faithful was fifty horses of a noble breed, fifty slaves, and 3000 pieces of gold. The feuds and disputes, which had detached Persia and Arabia from the caliphate of Syria, rendered the Ommiadan princes too feeble to oppose the pressing demands of the Christians; and the tribute was increased to a slave, a horse, and 1000 pieces of gold for each of the 365 days of the solar year, which continued to be paid until the empire was again united by the arms and policy of Abdolmalec.

In the reign of Soliman the Arabs made a second and equally unsuccessful attempt on Constantinople. The sound of war soon reached the Byzantine court; and while the most extensive preparations were making to repel the invaders, Moslemah, the brother of the caliph and governor of Upper Mesopotamia, was advancing at the head of 120,000 Arabs and Persians. In his progress through Asia Minor he reduced the cities of Tyana, Armoricum, and Pergamus; but before he reached the shores of the Hellespont, the Grecian sceptre had been transferred from Theodosius III. to the firmer hand of Leo the

Isaurian. At the well-known passage of Abydos the Mohammedan squadrons were transported from Asia to Europe; and by a circuitous march they invested Constantinople on the land side. Moslemah had instructed his troops to furnish themselves with two months' provisions, which they piled in a vast heap on one side of the encampment. They next planted their engines of assault, reared habitations of wood, and prepared the ground for tillage; thus declaring, by word and deed, their patient determination of expecting the return of seed-time and harvest, should the obstinacy of the besieged prove equal to their own.

The terrified citizens would willingly have purchased the departure of the invaders by an assessment of a piece of gold on the head of each inhabitant; but the proposal was rejected by Moslemah, whose confidence was elevated by the speedy approach of the Syrian and Egyptian navies, which are said to have amounted to 1800 ships. This huge armada made its appearance in the Bosphorus, and the Greeks beheld the smooth surface of the strait overshadowed with a moving forest. A night was fixed for a general assault by sea and land; and to allure the confidence of the assailants, Leo had caused the chain to be removed that usually guarded the entrance to the harbour. The stratagem took effect; for while the Saracens hesitated whether they should seize the opportunity or avoid the snare, the ministers of destruction were at hand. The Greeks introduced their celebrated fireships, and the Arabs, with their arms and vessels, were instantly wrapped in the unquenchable flames. Disorder seized their ranks; the flying barges were either dashed against each other or overwhelmed in the waves; and soon not a vestige remained of the fleet that had threatened to extirpate the Roman name.

Yet the siege was prolonged through the winter by

the neglect rather than the resolution of the new caliph. The season proved uncommonly rigorous; for more than 100 days the ground was covered with deep snow; and the natives of the sultry climes of Egypt and Arabia lay torpid and almost lifeless in their frozen tents. The return of spring revived their energies, and produced a second reinforcement of two numerous fleets, laden with corn, arms, and soldiers. But the Greek fires were again kindled; and if the destruction was less complete, it was owing to the experience which had taught the Moslems to remain at a safe distance; or to the perfidy of the Egyptians, who deserted to the service of the Christian emperor.

The calamities of famine and disease now began to make havoc among the ranks of the besiegers. After devouring every quadruped in their camp, they were reduced to such extremity as to depend for food, not only on the leaves and bark of trees, but on the most loathsome and disgusting substances. The spirit of conquest, and even of enthusiasm, became extinct. The Saracens, whenever they ventured to straggle beyond their lines, were exposed to the merciless retaliation of the Thracian peasantry. Twenty-two thousand of them were slaughtered by an army of Bulgarians, whom the gifts and promises of Leo had attracted from the banks of the Danube. To augment their distresses, a rumour was industriously spread that the Franks were arming by sea and land in defence of the Christian cause.

The siege had extended to thirteen months, when Moslemah was at length extricated from his difficulties by a letter from the Caliph Omar II., containing the welcome permission to retreat. The march of the Arabian cavalry through the Asiatic provinces was effected without hinderance or delay; but of the formidable host he had conducted to the gates of Constantinople not more than a sixth part returned to Damascus. The remains of the fleet were so damaged by the repeated attacks of tempest and fire, that only five galleys reached Alexandria.

The deliverance of Constantinople, in both sieges, must be ascribed, not to the want of prudence or courage on the part of the besiegers, but to the terrible efficacy of the Greek fire used on these occasions. In this extraordinary composition the principal ingredient was naphtha, or liquid bitumen, a tenacious and inflammable oil which springs from the earth, and catches fire as soon as it comes in contact with the air. This substance was mingled with certain proportions of sulphur, and pitch extracted from evergreen firs; and from the mixture, which produced a thick smoke and a loud explosion, proceeded a fierce flame, that burnt with equal vehemence in all directions, and was quickened, rather than extinguished, by the element of water.

This terrible compound obviously served the purpose of the rockets, bombs, and artillery of modern warfare; and might be employed by sea or land, in battles or in sieges. For the annoyance of the enemy, it was either poured from the ramparts in large boilers, or launched in red-hot balls of stone and iron, or darted in arrows and javelins twisted round with flax and tow which had deeply imbibed the inflammable oil. Sometimes, for a more sweeping destruction, it was deposited in fireships; and was most commonly blown through long tubes of copper, planted on the prow, and fancifully shaped into the mouths of savage monsters vomiting streams of consuming fire. The early French writers describe it as flying through the air like a winged long-tailed dragon, about the thickness of a hogshead, and dispelling the darkness of the night by its deadly illumination. The use of the Greek or maritime fire was afterward adopted by the Saracens, and continued to the middle of the fourteenth century, when the invention of gunpowder, an agent of similar but more powerful qualities, effected a new revolution in the military art.

Eastward, beyond Persia, the Ommiadan princes made considerable additions to their territories. In the caliphate of Othman, the Oxus formed the boundary of the Saracen empire. The rapacious conquerors repeatedly crossed that limit. Under Moawiyah the Tartars were driven into Bokhara; their queen, in her precipitate flight, leaving one of her slippers behind, which the Arabs valued at 2000 pieces of gold. These new acquisitions were confirmed and increased by Katibah, the lieutenant of Walid, who subdued the whole province of Khorasm, the district of Ferghana, and a part of Tartary, including the widely-extended regions between the Jaxartes and the Caspian. At the head of 20,000 men he proceeded to Samarcand, and invested the city, which had already made a nominal submission, but had taken a recent opportunity of violating the treaty. The haughty garrison, from their ramparts, taunted the besiegers with the vanity of those toils and dangers to which they were exposing themselves. The credulous Moslems, encouraged by an obscure prediction, redoubled their exertions, and by means of their warlike engines effected several breaches in the wall, which led to instant capitulation (A. D. 712). The inhabitants agreed to make an annual payment of 10,000,000 drachms (229,166*l.* 13*s.* 4*d.*), and a supply of 3000 slaves of the value of 200 drachms each (4*l.* 11*s.* 8*d.*), not one of whom was to be in a state of infancy, or ineffective from old age or debility. The ministers of religion were to be expelled from their temples, and their idols to be burnt or destroyed.

On entering the city, Katibah carried his stipulations into immediate execution. He performed the rites of Islam in the principal temple, which was converted into a mosque. The idols of pagan worship were collected into a heap, and with his own hand he set fire to the pile, which was soon consumed to ashes; and we are assured that 50,000 meskals of gold and silver were produced from the melted nails which had been used in the workmanship of the images. No unbeliever was to be allowed to remain within the gates longer than a ring of wet clay should preserve its moisture on his finger; for if it became dry before they quitted the place, every individual seen in arms, whether Turk or Tartar, was to suffer death without mercy. During a few years the citizens secretly cherished the idolatry of their ancestors; but the zeal and assiduity of their governor speedily effected their conversion (A. D. 728); and the Scythian shepherds at length abandoned the doctrines of Zoroaster, and the fire-worship of the Magians or Ghebers, for the faith of the Koran.

This remote city, however little it may now seem to claim the attention of civilized Europe, enjoyed at the time of its subjugation a comparative splendour. It was famous as the great resting-place of the caravans from China to Western Asia and Europe,—a source from which it could not fail to draw immense wealth. Silks and other luxuries were derived from the industry of the East, long before their manufacture was known by the Greeks or Romans; the principal route of the transit trade being across the Great Desert to Cashgar and Samarcand, and thence through Persia to Syria.

The first appearance of the star of Islam in Hindostan was about the forty-fourth year of the Hejira, when the province of Cabul had been reduced to subjection. Thirty years afterward, on the tributary impost being withheld, Obeidallah, the governor of Seistan, received orders from Hejaje to invade that country, and not to return till he had either subdued or totally destroyed the whole province. But he was deposed from the command and Abdurrahaman appointed to replace him, who with 40,000 men succeeded in conquering a large portion of that territory. A colony of Mussulmans settled among the mountains that extend between Moultan and Peshawer, where they employed themselves in the cultivation of the ground and the breeding of cattle; occasionally making war on the rajahs of the adjoining provinces. Though the Saracens had pushed their victories to the banks of the Indus, more than two centuries elapsed before a Mohammedan sovereign ruled in Lahore.

An embassy from Soliman proceeded to the capital of Kheten, beyond the Jaxartes, the residence of a Tartar prince (A. D. 715). Three successive days they approached the throne; the first, in plain white linen mantles, with slippers on their feet, and without uttering a word; the second, in habits and turbans of the richest silk, but still preserving the same inexplicable silence; the third, they made their appearance in complete armour, with scimitars by their sides, lances in their hands, their bows slung on their shoulders, and mounted on stately chargers. To the inquiries of the prince, who expressed his surprise at this frequent change of dress, Hobairah, the chief ambassador, replied, "On the first day our robes were those in which we visit our women and children; on the second, we appeared as we present ourselves at the court of our kings or governors; on the third, we wore the garb and equipment in which we always march against our enemies." Alarmed at this language, and at the daily accounts of their progress towards his frontiers, the prince courted the friendship and alliance of these fanatics, and dismissed the embassy with valuable presents.

When the Arabs first tried their valour at Muta against a foreign enemy, they could scarcely have anticipated that, before the close of a century, their empire should have exceeded in extent the greatest monarchies of ancient times; or that the successors of their Prophet should have risen to be the most powerful and absolute sovereigns on earth. Yet such was the fact. Their caliphs exercised a most unlimited and undefined prerogative, unfettered by popular rights, the votes of a senate, or the laws of a free constitution. They united in their own person the regal and sacerdotal characters; and if the Koran was the rule of their actions, they were the supreme judges and interpreters of that divine book. They reigned by the right of conquest over nations to whom the name of liberty was unknown, and who had not yet learned to detest those acts of violence and severity that were exercised at their expense. Under the last of the Ommiades, the Saracen empire extended 200 days' journey from east to west; and though the long and narrow province of Africa,—the sleeve of the robe, as their writers style it,—were withdrawn, the solid and compact dominion within the Jaxartes, the Hellespont, and the Indus, would spread on every side to the measure of five months of the march of a caravan in length, and four in breadth. From this estimate an important fragment was soon detached by the revolt of Spain; but its loss was more than counterbalanced by the subsequent conquests in India, Tartary, and European Turkey. This vast empire was ruled by a wretched political system, in which we seek in vain for the union and discipline that pervaded the government of Augustus and the Antonines. The only national feature was that general resemblance of manners and opinions which the progress of Islam had diffused over this immense space. The language and laws of the Koran were studied with equal devotion at Samarcand and Seville; the Moor and the Indian embraced as countrymen and brothers in the Temple of Mecca; and the Arabian language was adopted as the popular idiom in all the provinces to the westward of the Tigris.

END OF VOL I

www.ingramcontent.com/pod-product-compliance
Lightning Source LLC
LaVergne TN
LVHW011155110826
845150LV00006B/1229

* 9 7 8 1 4 2 5 5 4 6 2 8 1 *